PHYSIOLOGY DEMYSTIFIED

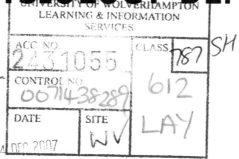
The Hon. Dr. Dale Pierre Layman, Ph.D., Grand Ph.D. in Medicine (Belgium)

McGRAW-HILL

New York   Chicago   San Francisco   Lisbon   London
Madrid   Mexico City   Milan   New Delhi   San Juan
Seoul   Singapore   Sydney   Toronto

The **McGraw·Hill** Companies

Cataloging-in-Publication Data is on file with the Library of Congress

Copyright © 2004 by The McGraw-Hill Companies, Inc. All rights reserved. Printed in the United States of America. Except as permitted under the United States Copyright Act of 1976, no part of this publication may be reproduced or distributed in any form or by any means, or stored in a data base or retrieval system, without the prior written permission of the publisher.

4 5 6 7 8 9 0   DOC/DOC   0 1 0 9 8 7 6

ISBN 0-07-143828-9

The sponsoring editor for this book was Judy Bass and the production supervisor was Pamela A. Pelton. It was set in Times Roman by Keyword Publishing Services, Ltd. The art director for the cover was Margaret Webster-Shapiro; the cover designer was Handel Low.

Printed and bound by RR Donnelley.

McGraw-Hill books are available at special quantity discounts to use as premiums and sales promotions, or for use in corporate training programs. For more information, please write to the Director of Special Sales, McGraw-Hill Professional, Two Penn Plaza, New York, NY 10121-2298. Or contact your local bookstore.

PHYSIOLOGY DEMYSTIFIED

Demystified Series

This book is fondly dedicated to my wonderful wife, Kathy. She is the "Anchor," the solid "Rock" upon which we can all lean in times of trouble or need. She is my best and most patient friend.

I also wish to highlight Allison, one of my artistically talented daughters. It is through her creative efforts that our host – Professor Joe, the Talking Skeleton – has been brought to vibrant life.

Finally, I wish to thank Janet M. Evans, President of the American Biographical Institute, and Nicholas S. Law, Director General of the International Biographical Centre (Cambridge, England). In many of their volumes, they have described my ideas for "Intuitive Geometry and the A&P (Human Anatomy & Physiology) Text."

And, now, through the help of Judy Bass and Scott Grillo at McGraw-Hill, these ideas are being realized!

CONTENTS

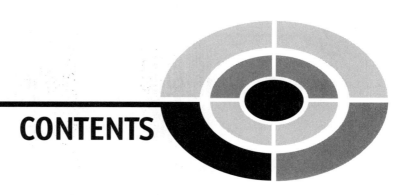

CONTENTS

PREFACE

"Which comes first – the chicken, or the egg?" This book is about the "chicken" – *human physiology* or living body *functions*. (The process of laying an egg is an aspect of chicken physiology, after all!) Its close companion, *ANATOMY DEMYSTIFIED*, is all about the "egg" – *human anatomy* or body *structure*. (An unhatched egg is an example of anatomy.)

PHYSIOLOGY DEMYSTIFIED is for people who want to get acquainted with the fundamental concepts of human body function, without having to take a formal course. But it can also serve as a supplemental text in a classroom, tutored, or home-schooling environment. In addition, it should be useful for career changers who need to refresh their knowledge of the subject. I recommend that you start at the beginning of this book and work straight through.

This book seeks to provide you with an intuitive, highly visual grasp of physiology and its terminology. Starting with the Great Body Pyramid (a concept borrowed from anatomy and the Ancient Egyptians), *PHYSIOLOGY DEMYSTIFIED* guides you along *A Living Path Through Bodyspace*. Professor Joe, the Talking Skeleton, is our host. He is drawn as a cartoon standing upright and pointing, whenever key facts about Biological *Order* in the human body are being presented. But when he is fallen and fractured, our Good Professor is talking to you about facts of Biological *Disorder* within the human body. These key facts of Order-versus-Disorder will be about Anatomy, Physiology, or just Plain Body Functions.

As you go from body system to body system, you will also learn *where* to put many facts of Biological Order/Disorder, briefly writing them within the "grids or drawers" of the Great Body Pyramid. In this way, like putting your socks into a drawer, you will always know where to find the key body facts ("socks") whenever you need them!

This introductory work also contains an abundance of practice quiz, test, and exam questions. They are all multiple-choice, and are similar to the sorts of questions used in standardized tests. There is a short quiz at the end of every chapter. The quizzes are "open-book." You may (and should) refer to the chapter texts when taking them. When you think you're ready, take the quiz, write down your answers, and then give your list of answers to a friend. Have the friend tell you your score, but not which questions you got wrong. The answers are listed in the back of the book. Stick with the chapter until you get most of the answers correct.

This book is divided into six sections. At the end of each section is a multiple choice test. Take these tests when you're done with the respective sections and have taken all the chapter quizzes. The section tests are "closed-book," but the questions are not as difficult as those in the quizzes. A satisfactory score is three-quarters of the answers correct. Again, answers are in the back of the book.

There is a final exam at the end of this course. It contains questions drawn uniformly from all the chapters in the book. Take it when you have finished all six section tests, and all of the chapter quizzes. A satisfactory score is at least 75 percent correct answers.

With the section tests and the final exam, as with the quizzes, have a friend tell you your score without letting you know which questions you missed. That way, you will not subconsciously memorize the answers. You can check to see where your knowledge is strong, and where it is not.

I recommend that you complete one chapter a week. An hour or two daily ought to be enough time for this. When you're done with the course, you can use this book, with its comprehensive index, as a permanent reference. What you now hold in your hand, I think you will agree, is a most *unusual* approach to the study of human physiology! We have, of course, our most unique talking skeleton host, Professor Joe (and occasional glimpses of his somewhat mischievous sidekick, Baby Heinie). More importantly, this book represents the practical application of what I like to call *Compu-think*, or "*compu*ter-like modes or ways of human *think*ing." This is reflected in its heavy emphasis upon binary (two-way) classifications, grid-associated reasoning, and frequent occurrence of summary word equations.

Suggestions for future editions are welcome.

Now, work hard! But, be sure to have *fun*! Best wishes for your success.

The Hon. Dr. Dale Pierre Layman, Ph.D., Grand Ph.D. in Medicine

ACKNOWLEDGMENTS

The most interesting and entertaining illustrations in this book are mainly due to the talented efforts of one of my own daughters, Allison Victoria Layman. It is through her gifted eyes that my visions for picturing key body concepts have been successfully brought to life!

I extend thanks to Emma Previato of Boston University, who helped with the technical editing of the manuscript of this book. Thanks also go to Maureen Allen and the staff at Keyword, who helped me winnow out various errors and inconsistencies within the original manuscript.

I particularly wish to thank Judy Bass, Senior Acquisitions Editor at McGraw-Hill. Judy has been *very* enthusiastic and supportive of all my writing efforts! This means *a lot* to a struggling author! She also deserves credit for her brilliant insight that we need two separate but closely linked books – an *ANATOMY DEMYSTIFIED*, as well as a stand-alone *PHYSIOLOGY DEMYSTIFIED* – in this series.

Finally, Mr. Scott Grillo (publisher) has been a quiet, steadfast, and kindly presence behind all of my writing efforts. I am most pleased to honor both Judy Bass and Scott Grillo within the distinguished pages of the *Dedication Section* in *2000 OUTSTANDING INTELLECTUALS OF THE 21ST CENTURY* (2nd Ed., 2003), just published by the International Biographical Centre (Cambridge, England). The unusual and creative thinking efforts being presented in these humble little volumes are closely being watched and reported on, within the Highest World Intellectual Circles!

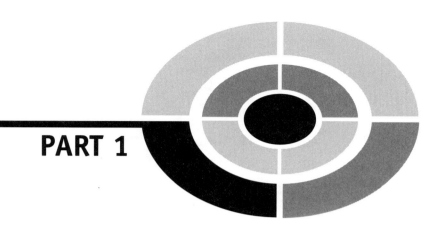

PART 1

The Golden Path of Body Function

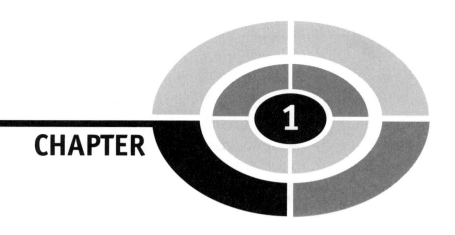

CHAPTER 1

Physiology: Our Life Is A Path Through Bodyspace

Welcome! Welcome to PHYSIOLOGY DEMYSTIFIED: A Living Path Through Bodyspace

Hello, there! Who am *I*? Why, I am your host. They call me Professor Joe, the Talking Skeleton! I have been selected as your guide for this book, *PHYSIOLOGY DEMYSTIFIED*. I am here to give you a basic, "bare bones" introduction to *what happens* in The Place Below Your Skin! You and I are about to take a fascinating walk through the living body. We will describe this walk using a colorful phrase . . . *A Living Path Through Bodyspace*.

All Life Starts with Biology

Before we explore *physiology* (pronounced as **fih**-zee-**AHL**-uh-jee), we need to review the broader subject of *biology* (buy-**AHL**-oh-jee). The word biology is really a technical term that arises from two word parts of Ancient Greek – *bi* ("life") plus *-ology* ("study of").

Thus biology literally means the "study of life." (A related book, *BIOLOGY DEMYSTIFIED*, covers this broad subject of general biology in considerable detail.) Since our subject of interest is *human* physiology, we will be focusing upon the part of general biology that includes it.

BIOLOGICAL ORDER = "LIVING PATTERNS"

Soon after you begin studying biology (or physiology), you become conscious of its many distinct *patterns*. In general, a pattern is some particular arrangement of shapes, forms, colors, or designs. [**Study suggestion:** Closely examine some of the patterns you find outside and around your own home or apartment, such as its distinct grid or map of intersecting streets, various buildings, and sidewalks.]

For our treatment of physiology, however, we will be concentrating upon various patterns within the human *organism* (**OR**-gan-**izm**) – the entire living human body. Speaking broadly, an organism is any living body with a high degree of *Biological Order*. By Biological Order, we simply mean a recognizable pattern involving one or more organisms.

[**Study suggestion:** Look up into the sky and let your eyes trace the pattern made by a bird as it flies overhead. Does this pattern of flight represent a case of Biological Order? How does this type of order differ from that found in the specific arrangement of streets, buildings, and sidewalks around your home?]

Specifically, we want to study the patterns of Biological Order associated with the human body. Since the human body is a living organism, it shares with other living creatures this fundamental relationship:

| BIOLOGICAL ORDER | — | "LIVING" PATTERNS (PATTERNS IN LIVING OR ONCE-LIVING ORGANISMS) |

Figure 1.1, below, provides a number of distinct patterns. Look them over. Which of the patterns probably represent *Biological* Order of the *human organism*?

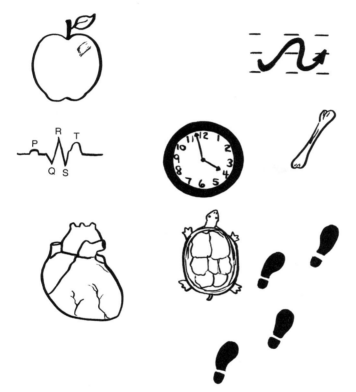

Fig. 1.1   Some distinct patterns of order.

BIOLOGICAL DISORDER = A "BREAK IN LIVING PATTERNS"

Now that we have discussed the notion of Biological Order and seen some specific examples of it, it is time to consider its exact *opposite*! If Biological

Order represents *intact* patterns, then *Biological Disorder*, of course, would represent the *breaking* or *absence* of such patterns:

Take a quick peek back at Figure 1.1 and its orderly patterns. Try to picture how *breaking* these patterns of order, thereby creating states of *disorder*, would look. Now, follow this up with an examination of Figure 1.2. This time, ask yourself, "Which of the broken patterns probably represent *Biological* Disorder of the *human organism*?"

Fig. 1.2 Some broken patterns of order.

ICONS FOR BIOLOGICAL ORDER VERSUS DISORDER

Special *icons* (**EYE**-kahns) or symbols are used in this book to tag key body facts as being either examples of Biological Order (an intact Professor Joe, the Talking Skeleton) or Biological Disorder (a fallen and fractured Professor Joe). (Consult Figure 1.3.)

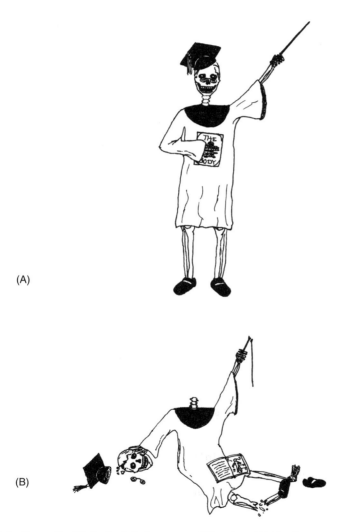

(A)

(B)

Fig. 1.3 Icons for Biological Order versus Disorder. (A) Professor Joe standing upright: an icon for biological order. (B) Professor Joe fallen and fractured: an icon for biological disorder.

Two Paths Curve Apart in the Woods

We have said that we are on a nice walk, a nice walk through the Woods of the Human Body. But soon after we begin our exciting journey, we see two different paths quickly splitting off, and curving away into the darkness. One of these paths we shall call *anatomy* (**ah-NAT**-oh-me), while the other will be known as *physiology*. The splitting of these two paths of human body study – anatomy from physiology – formally began during the 1500s, and their separation has been getting wider ever since!

BODY *STRUCTURE* BECOMES *ANATOMY*

The word anatomy derives from Ancient Greek. It literally translates to mean "the process of" (-*y*) "cutting" (*tom*) "up or apart" (*ana*-). This exact translation of anatomy clearly reflects its close connection to *dissection* (dih-**SEK**-shun), which has essentially the same meaning. In each case, the thing being dissected is the human body and its structures. Anatomy, therefore, can be defined as body structure and the study of body structures, primarily by means of dissection. (A close companion volume to this book, *ANATOMY DEMYSTIFIED*, looks at the various topics in human anatomy in much greater detail.)

Body structures would include such things as bones, muscles, the heart, and brain. Considered from a common grammar standpoint, each of these body structures can be considered a noun in a sentence. This is because a noun is a person, place, or thing. In the specific case of anatomy, then, the body structures are important parts of a person (human organism).

Aristotle, the Father of Natural History

One very important name for you to know in the history of both anatomy and physiology is *Aristotle* (**AIR**-ist-ahtl). Living between 384 and 322 B.C. in Ancient Greece, Aristotle is often considered the *Father of Natural History* – the collection and classification of plants and animals into certain groups. Aristotle is also often considered the world's first great biologist. Aristotle gathered huge amounts of information about both the anatomy and the physiology of numerous creatures in Nature.

Human anatomy = our inner world of "bodyspace"

Aristotle looked at spaces – both the outer space of the Universe and the inner space within humans. For convenience, let us call our inner world by the term *bodyspace*.

To Aristotle and many other early Greek thinkers, outer space and its "heavenly bodies" (such as the sun, moon, and stars) made up a *macrocosm* (**MAK**-ruh-**kahz**-um). This macrocosm was a very "large" (*macr*) "universe" (*cosm*). Conversely, the much smaller human body represented a *microcosm* (**MY**-kruh-**kahz**-um) or "tiny" (*micr*) "universe."

To Aristotle, then, the *internal* (in-**TER**-nal) or "inner" environment deep within the human body was merely a tiny chunk of a much grander *external* (**EKS**-ter-nal) or "outer" environment that surrounded it. The macrocosm (external environment) thus reached far into outer space and its heavenly bodies. But the microcosm (internal environment) started at the surface of the human skin, and only included the space deep within the body. (These differences are clearly illustrated within Figure 1.4.)

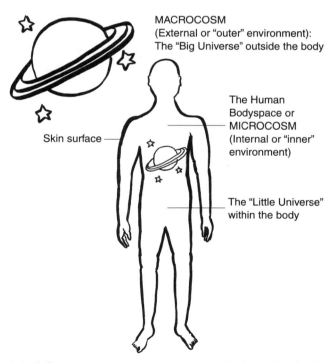

Fig. 1.4   Microcosm versus macrocosm: two "universes" of order.

Human bodyspace: a series of stacked grids within a Pyramid

One highly orderly model for human bodyspace (the internal environment) is a series of horizontal grids, stacked one upon the other to create a Great Pyramid (Figure 1.5).

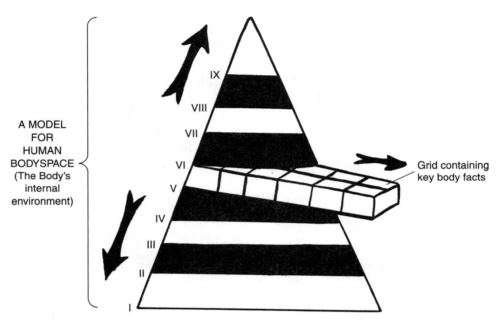

A MODEL
FOR
HUMAN
BODYSPACE
(The Body's
internal
environment)

IX
VIII
VII
VI
V
IV
III
II
I

Grid containing
key body facts

Fig. 1.5   Human bodyspace as a pyramid of stacked grids.

LIVING BODY *FUNCTIONS* BECOME *PHYSIOLOGY*

Having explained the basics of anatomy (body structure) and human bodyspace (the internal environment), it is now time to talk about *what goes on* within this orderly space! The other pathway we now follow is that of body *function* – some type of body "performance." A function is something that a structure does, or something that is done to a structure. Consider, for instance, the following statement: "The hammer hit the nail." Both the hammer and the nail are structures (exist as nouns), while *hit* represents a function. Such functions, therefore, can generally be considered as action verbs within sentences or phrases.

The original translation of *physiology*

When one goes far back in time, returning again to Ancient Greece, we learn that Aristotle was the person who actually created the word *physiology*. If we mentally "dissect" physiology by inserting slashmarks into it, we can discover what Aristotle was thinking (Figure 1.6).

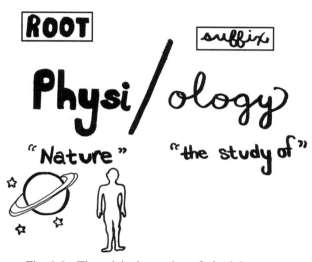

Fig. 1.6   The original meaning of physiology.

To Aristotle, physiology was literally "the study of" (*-ology*) "Nature" (*physi*). By this interpretation, physiology involved studying practically *everything* in the Universe *and* in the human body, because *both* of them made up The World of "Nature"! Thus physiology originally meant both the macrocosm (external environment) and the microcosm (internal environment of bodyspace). We have already learned that both the internal and external environments were considered by the early Greeks to be reflections of the same basic thing – the existence of *physi* (**FIH**-zee), which to them translated to mean "law-following patterns of order," as well as "Nature."

Galen begins modern experimental physiology

Even though it was Aristotle who first coined the term physiology, another man is usually given credit for changing the interpretation of the word into its modern usage. This man was named *Claudius Galen* (**GAY**-lun), who lived in

Greece and Rome during A.D. 130–200. Galen, a philosopher and physician, is often called *the Father of Experimental Physiology*. This is because Galen was among the first to perform *experiments* – controlled trials – on *living* animals such as dogs, pigs, bears, and apes. The technical term for what he did is *vivisection* (**viv**-uh-**SEK**-shun), the "process of cutting" (*section*) "living" (*vivi*) things apart.

Galen's thousands of vivisections of living animals (rather than merely dead humans or animals) brought their true body functions to light. Thus, the modern science of physiology was born! In its modern sense, *physiology is the study of **living** body **functions***, that is, it is the study of the **nature** of **living** things, **only**.

Galen made this important point quite clear when he operated on living animals and cut open their hearts and blood vessels. When Galen plunged a knife into the *left ventricle* (**VEN**-trih-**kl**), a "little belly"-like cavity at the bottom of an animal heart, blood came spurting out of it (Figure 1.7). Further, Galen showed that the *arteries* (**AR**-ter-**ees**), which were wrongly thought to be "air keepers," also contained flowing blood. By such experi-

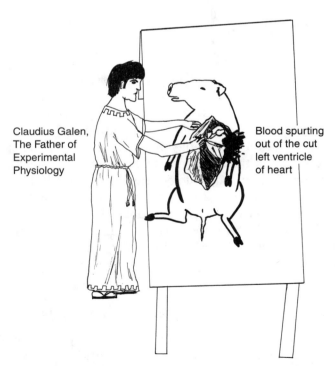

Claudius Galen, The Father of Experimental Physiology

Blood spurting out of the cut left ventricle of heart

Fig. 1.7 Galen demonstrates the presence of blood within the heart.

ments on living animals, then, Galen showed that modern physiology is restricted to the study of functions of the living body.

The living path: a curving path that passes *through* grids

We earlier modeled anatomy (body structure) as a series of stacked grids within a Body Pyramid. Physiology can thus be viewed as a curving "Pathway of Life" that passes through these grids of Bodyspace.

Contrasting Physiology with Plain Functions

In addition to contrasting physiology from anatomy, it is also important to be able to distinguish physiology from a number of *plain functions* that occur within the body. Plain functions are the actions that the various *non-living* structures in the body perform. Consider, for instance, the *carbon (C) atom.* The billions of carbon atoms in the human body tend to form *chemical bonds* with each other. The result is long strings of bonded carbon atoms. Since carbon atoms are, of course, non-living, their bonding together is an example of a plain body function, rather than physiology.

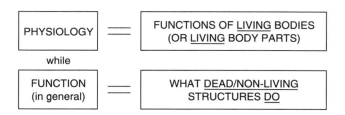

| PHYSIOLOGY | —— | FUNCTIONS OF <u>LIVING</u> BODIES
(OR <u>LIVING</u> BODY PARTS) |

while

| FUNCTION
(in general) | —— | WHAT <u>DEAD/NON-LIVING</u>
STRUCTURES <u>DO</u> |

Letter Symbols for Anatomy, Physiology, and Functions

We have already classified key body facts as either being examples of Biological Order (symbolized by an intact Professor Joe), or examples of Biological Disorder (a fallen and fractured Professor Joe). Now, we will also classify them according to whether they represent anatomy (designated by a black capital **A**), physiology (symbolized by a white capital **P**), or plain function (tagged by a white capital **F**).

The resulting icons representing all possible combinations of Order/ Disorder with **A**, **P**, or **F**, are shown in Figure 1.8.

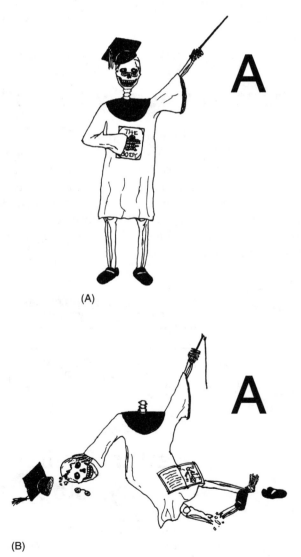

(A)

(B)

Fig. 1.8   Icons representing key facts with various combinations.

(C)

(D)

Fig. 1.8 (continued)

(E)

(F)

Fig. 1.8 (continued)

Some Characteristics of Living Body Functions

Since we are distinguishing facts of physiology (*living* body functions) from facts of plain body function (*non-living* body actions), it is essential for us to describe some of the characteristics of living body functions:

1. Living body functions tend to remain relatively constant over time. The resulting pattern of Biological Order is usually called *homeostasis*. For example, a living human organism maintains an *oral* (**OR**-al) body temperature taken by "mouth" (*or*) at a relatively constant level of about 98.6 degrees Fahrenheit. (Examine the diagram in Figure 1.9.) To be sure, the oral body temperature does tend to rise and fall over time. Nevertheless, it still stays within a relatively narrow band, called the *normal range*. The temperature tends to rise toward an *upper normal limit* or ceiling value of about 99.6 degrees F. Conversely, the temperature tends to fall toward a *lower normal limit* or floor value of about 97.6 degrees F.

Note from Figure 1.9 that oral body temperature is not *absolutely* constant over time, since it does rise and fall. Rather, it is best described as being *relatively constant* over time – never rising above its normal range, nor falling below it. We generally call this roughly S-shaped pattern of relative constancy a state of *homeostasis* (**hoh**-mee-oh-**STAY**-sis).

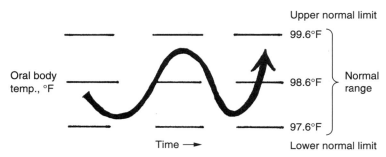

Fig 1.9   Homeostasis of oral body temperature over time.

2. Living body functions are usually *complementary* to particular body structures. Two things are considered *complementary* (**kahm**-pluh-**MEN**-tuh-ree) when, taken together, they make a meaningful whole. According to the *Law of Complementarity* (**kahm**-pluh-men-**TAIR**-uh-tee), anatomy or body structure determines physiology. By this it is meant that a particular body structure, by its very composition and basic characteristics, can only perform certain body functions.

Consider, for instance, the *femur* (**FEE**-mur), a long bone of the "thigh" (*femor*), as well as the human eye. As is plain from an examination of Figure 1.10, the femur has the basic structural characteristics of a long, white, rigid pillar. And the interior of the human eyeball consists of a complex series of lenses and fluid-filled compartments. This makes the eyeball somewhat like a complicated telescope or camera in its structure.

The pillar-like structure of the femur, therefore, "determines" its major body function or physiology – supporting the body's weight. And the camera-like anatomy of the eyeball "determines" its major function of focusing light rays for the physiology of vision.

[**Study suggestion:** Pretend that you have plucked both of your eyeballs out of their sockets, and that you have amputated both of your lower legs just above each femur. Now, jam your amputated femurs into your empty

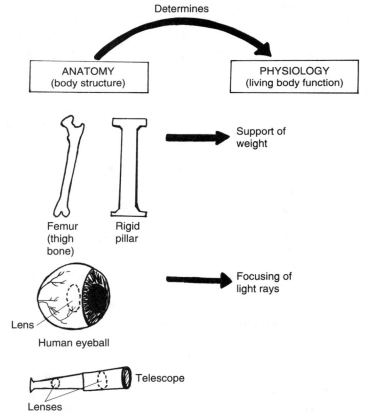

Fig. 1.10 The Law of Complementary between body structure and function.

eye sockets, and strap both of your eyeballs to your leg stumps! Next, lower your body onto the floor, and try to walk out of the room. Well, did you make it? How does the Law of Complementarity make this situation so impossible?]

3. Living body functions generally consume energy during metabolism, and give off heat in the process. If a particular part of the body is alive, then it is always engaging in *metabolism* (meh-**TAB**-ah-**lizm**) or a "state of change." Food that is eaten is soon changed by the chemical processes of metabolism. Energy is produced, which then performs body work. Such body work is usually some aspect of physiology, such as moving the parts of the body around. And in this process of consuming energy and doing work, a considerable amount of heat is produced. (In other words, living things are usually a lot hotter than dead ones!)

4. Living body functions are usually sensitive to changes that occur within the internal or external environment, and they respond to these changes. A *stimulus* (**STIM**-you-**lus**) is literally a "prod" or "goad." (Picture a long stick that pokes or prods the body.) In general, a stimulus is a detectable change in the body's internal or external environment. "What detects this change?" the curious reader may well ask. The answer is: a *sensory receptor* (ree-**SEP**-ter). A sensory receptor is a group of specialized cells or modified nerve endings that "receive" (*recept*) the stimulus, and are excited or aroused by it. Consider, for example, a rise in oral body temperature towards its upper normal limit of 99.6 degrees F. This rise is detected by groups of *thermo-receptors* (**ther**-moh-ree-**SEP**-ters) in the skin. The thermoreceptors are actually naked nerve endings that are very sensitive to changes in body "heat" (*therm*).

The thermoreceptors alter their physiology by firing off *nerve impulses*. These nerve impulses, in turn, activate certain *body effectors* (e-**FEK**-ters). An effector is part of the body that carries out a particular *response*, thereby having some "effect" upon the environment. The sweat glands in the skin, for instance, are body effectors that respond to stimulation via nerve impulses by increasing their rate of sweat secretion. As a result, the oral body temperature soon declines.

In general, we see the typical *physiological* (**fih**-zee-oh-**LAHJ**-uh-kul) sequence of stimulus, sensory receptors, body effectors, and response:

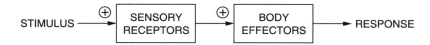

5. Whenever living body functions become highly irregular and disordered, a state of pathophysiology and disease often results. A quick review of Figures 1.1 and 1.2 provides contrasting examples of Biological Order versus Disorder. In the second row of illustrations in Figure 1.1, the series of waves shown on the left represents a typical human *electrocardiogram* (ih-**lek**-troh-**KAR**-dee-oh-gram). An electrocardiogram is literally a "graphical record" (-*gram*) of the "electrical activity" (*electro*-) of the "heart" (*cardi*). The electrocardiogram is commonly abbreviated as either *ECG* or *EKG*. The normal EKG (ECG) shows an orderly pattern starting with a *P wave*, which leads to a *QRS wave sequence*, and is finally followed by a large *T wave*. This normal EKG pattern is typical of healthy *cardiac* (**KAR**-dee-**ak**) – "pertaining to" (-*ac*) the "heart" (*cardi*) – *physiology*.

The second row in Figure 1.2, in great contrast, shows an EKG pattern with a high degree of *Biological Disorder* or *electrical chaos* (**KAY**-ahs). The waves of this EKG are all scrambled and unrecognizable. The result is a bad case of *cardiac pathophysiology* (**path**-oh-**fizz**-ee-**AHL**-uh-jee). In general, pathophysiology represents a "diseased" (*path*) and highly abnormal state of physiology. For the patient with this scrambled EKG, the heart may well stop pumping, resulting in *cardiac standstill*!

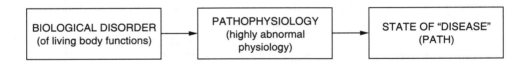

6. Living body functions are associated with one or more *Levels of Body Organization*. (This topic will be discussed in detail within the next section.)

The Different *Levels* of Physiology

Earlier in this chapter, we talked about the concept of anatomy being modeled in a particular way. We thought of body structures as things occupying Human Bodyspace. And the internal environment was modeled by a series of stacked grids within a Great Body Pyramid. We are now ready to extend this concept to include physiology, because of its close relationship with anatomy. Recall that, according to the Law of Complementarity, anatomy or body structure determines physiology.

THE GREAT PYRAMID OF STRUCTURE-FUNCTION *ORDER*: *"NORMAL"* BODY PATTERNS

Every body structure in the anatomy of a living human organism essentially *does* something! That is, every body structure performs a certain aspect of either plain function or physiology. Therefore, it is appropriate to talk about Body Structure-Function Pyramids. Let us picture *the Great Pyramid of Structure-Function **Order***. This Great Pyramid consists of *"Normal" Body Patterns*, each of them occurring at a particular *Level of Body Organization*.

A Level of Body Organization is a particular degree of size and complexity of body structures. In general, we can picture the Great Pyramid of Structure-Function Order as consisting of nine different levels of body organization (Figure 1.11, A).

Near the base of the Pyramid, the body structures of anatomy are simpler and smaller. But as we climb toward the peak or apex of the Pyramid, the body structures become progressively larger and more complex.

Figure 1.11 (B) gives specific names to each of the nine levels of body organization in the Pyramid. Starting at the base, these are called the *subatomic* (sub-ah-**TAH**-mik) *particles*, *atoms*, *molecules*, *organelles* (**OR**-gah-**nels**), *cells*, *tissues*, *organs*, *organ systems*, and the entire human *organism*.

THE GREAT PYRAMID OF STRUCTURE-FUNCTION *DISORDER*: *"ABNORMAL"* BODY PATTERNS

"What about body structures and functions that are in a state of Biological *Disorder*?" the inquiring reader needs to question. For a reply, one need only look at Figure 1.11 (C). Here we picture *the Great Pyramid of Structure-Function **Disorder***. Shown alongside this Pyramid is our old friend, Professor Joe, in a fractured and fallen-down state. Once again, Levels I–IX (subatomic particles through the organism) are shown. But here the body structures and functions associated with these levels are in a state of disorder, such that they represent "abnormal" body patterns. Their anatomy can be called *pathological* (**path**-oh-**LAHJ**-ih-**kal**) *anatomy*, while their physiology can likewise be called pathophysiology. In short, both pathological anatomy and pathophysiology are similar in that they generally reflect "disease" (*path*).

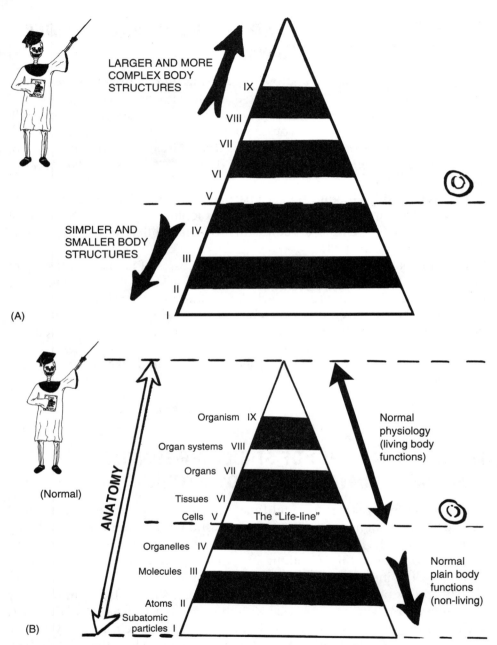

LARGER AND MORE COMPLEX BODY STRUCTURES

SIMPLER AND SMALLER BODY STRUCTURES

(A)

Organism IX
Organ systems VIII
Organs VII
Tissues VI
Cells V
Organelles IV
Molecules III
Atoms II
Subatomic particles I

The "Life-line"

(Normal)

ANATOMY

Normal physiology (living body functions)

Normal plain body functions (non-living)

(B)

Fig. 1.11 The Great Pyramids of body structure and function. (A) The Great Pyramid of Structure-Function Order. (B) Specific Features associated with the Great Pyramid of Structure-Function Order. (C) The Great Pyramid of Structure-Function Disorder.

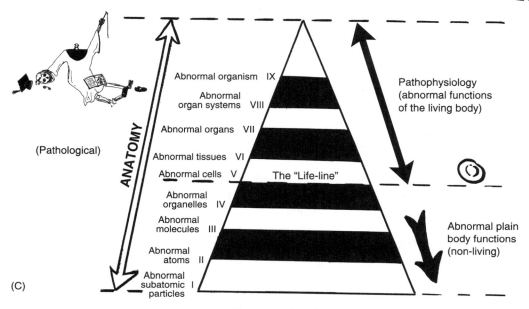

Fig. 1.11   (continued)

THE BODY "LIFE-LINE": PHYSIOLOGY BEGINS AT THE CELL LEVEL

A quick glance back at both Figures 1.11 (B) and (C) reveals the presence of a *"Life-line"* in each Great Pyramid. This "Life-line" is the *Cell Level* of Body Organization (Level V) within the stacked grids of each Pyramid. *The reason is that the cell represents the lowest **living** level of Body Organization. Hence, **physiology** only occurs from the Cell Level on up to the Organism Level. Below the Cell Level (organelles down to subatomic particles), only **anatomy** and **plain function** exist.* (The physiology of the cell is discussed in Chapter 4.)

Summary: A Three-Way System of Classification for Key Body Facts

As we approach the end of this first chapter, we have laid the foundations for creating a *three-way system of fact classification:*

1. Classification according to Biological Order/Disorder. A key body fact can be labeled with Professor Joe standing and pointing if it represents a case of Biological Order. If it is a case of Biological Disorder, however, then it is labeled with an icon of Professor Joe fallen and fractured.

2. Classification according to identity as either Anatomy, Physiology, or Plain Body Function. If the key text fact being highlighted represents Anatomy, then a black capital **A** is placed under Professor Joe's pointer (either intact or broken). If the fact is Physiology, then a white capital **P** is placed under the intact/broken pointer. Finally, if the fact is just plain Function, a white capital **F** is placed under the Professor's pointer.

3. Classification according to Level of Body Organization. The third way of classifying a key fact is according to the level of body organization that it represents. Consider, for example, a key fact such as this one: "The normal resting heart usually beats at a rate of about 72 times per minute."

Organ 1

First, the standing Professor Joe is used, because the heart is in a state of Biological Order. Under the Professor's pointer, we insert a white **P**, because an aspect of normal physiology of the heart (its rate of beating) is occurring. And thirdly, we have the word **organ**, written under the **P**, because it is the physiology of the entire heart organ that is being described. (See Figure 1.12, A.)

Organ 1

Now consider this sentence: "About 2/3 of the heart is located on the left side of the body midline, in most individuals." The black capital **A** is used under standing Professor Joe for normal anatomy and a state of Biological Order. The word **organ** is written under the **A**, because it is the anatomy or structure of the entire heart that is being discussed. (See Figure 1.12, B.)

INSERTION WITHIN A BODY-LEVEL FACT GRID

At the end of each chapter, a number of *Body-Level Fact Grids* are provided.

Think of each grid as an open drawer or matrix of square cells within the Great Body Pyramid:

1 Key Fact #1	2 Key Fact #2
3 Key Fact #3	4 Key Fact #4

Some particular Level of Body Organization (I, subatomic particles → IX, organism)

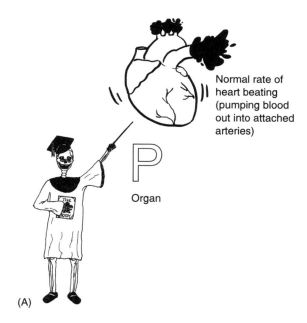

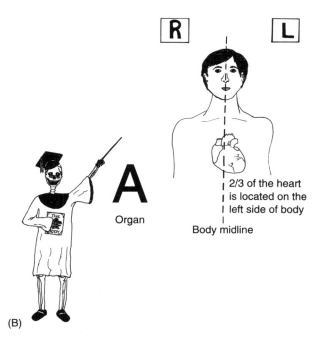

Fig. 1.12   An example of a three-way classification of a key body fact. (A) An example of normal physiology, Biological Order, at the organ level. (B) An example of normal anatomy, Biological Order, at the organ level.

Each of the four numbered cells or compartments within each grid provides a specific place for you to briefly write-in and summarize the key facts you have read, about a particular body structure and, quite often, its associated physiology. Since the grids essentially represent levels of body structure (anatomy), the frequent placing of physiological facts into the grids suggests that physiology is weaving A Golden Path – A *Living* Pathway Through Bodyspace! (This colorful metaphor was presented at the beginning of this chapter.)

As you progress through *PHYSIOLOGY DEMYSTIFIED*, chapter by chapter, you can conveniently return to these Body-Level Grids and review and retrieve the information stored there. In this way, you will be learning to think much like an Ancient Egyptian, who liked to put almost everything somewhere into a Pyramid!

Quiz

Refer to the text in this chapter if necessary. A good score is at least 8 correct answers out of these 10 questions. The answers are listed in the back of this book.

1. Biology and physiology have in common:
    (a) Their concentration upon the structure of living plants and animals
    (b) A primary focus upon living body functions
    (c) An avoidance of anything having to do with diseases
    (d) Study of dead people, plants, and animals

2. Biological Order can best be described as:
    (a) The study of the nature of living things
    (b) Essentially the same thing as randomness and chaos
    (c) The occurrence of various patterns within living organisms
    (d) Having no significant relationship to either anatomy or physiology

3. The word anatomy is closest in its literal translation to the word:
    (a) Dissection
    (b) Physiology
    (c) Nature
    (d) Function

4. The Father of Natural History is:
    (a) Jules Verne
    (b) Baby Heinie
    (c) Hippocrates
    (d) Aristotle

5. The Ancient Greek idea of a Macrocosm was basically equivalent to:
    (a) The internal environment
    (b) Homeostasis
    (c) The normal range
    (d) The external environment

6. "The kite flew over the roof of the barn." In this sentence, the word *flew* represents:
    (a) Physiology
    (b) Anatomy
    (c) Plain function
    (d) A noun

7. Claudius Galen is usually considered the:
    (a) Inventor of the microscope
    (b) Creator of the word physiology
    (c) First person to systematically dissect human cadavers
    (d) The Father of Experimental Physiology

8. The sodium (Na) atoms in the human body fluids can form chemical bonds with the chlorine (Cl) atoms. This bonding process is a case of:
    (a) Plain body function
    (b) Biological Disorder
    (c) Physiology
    (d) Metabolism

9. Relative constancy of blood glucose (sugar) concentration over a 24-hour period:
    (a) Upper normal limit
    (b) Homeostasis
    (c) Heterostasis
    (d) Sensory reception

10. A good example of the Law of Complementarity:
    (a) Lungfish crawling out onto the land to search for food
    (b) Human red blood cells exploding when placed into distilled water
    (c) A gentlemen tipping his hat to a lady passing on the sidewalk
    (d) Teeth of sharks being put onto necklaces by South Seas natives

Body-Level Grids for Chapter 1

Several key body facts were tagged with numbered icons in the page margins of this chapter. Write a short summary of each of these key facts into a numbered cell or box within the appropriate *Body-Level Grid* that appears below.

Anatomy and Biological Order **Fact Grid for Chapter 1:**

A

ORGAN
Level

1

Physiology and Biological Order Fact Grid for Chapter 1:

ORGAN
Level

1

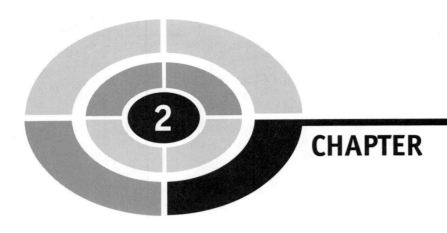

CHAPTER 2

Control of the Internal Environment: A Story About Feedback and Homeostasis

We have now begun our Magical Journey down a *Living* Path – Human Physiology, our Living Path through the World of "Bodyspace."

Chapter 1 taught us to look for normal patterns of Biological Order within the Great Pyramid of Structure-Function Order.

Enter Hippocrates: Human Health Seen as a Balance

Closely associated with patterns of order is the notion of *balance*. A balance, in general, is a rough equality between two or more different things. Consider, for example, the physiological concept of "eating a balanced diet." According to this view, to maintain their long-term health and well-being, people should not over-consume any one particular type of food in their diet. Rather, they should eat all three major types of foodstuffs – *proteins*, *lipids* (**LIP**-ids), and *carbohydrates* (**kar**-boh-**HIGH-draytes**) – in roughly equal or "balanced" amounts.

This hearkens back to the time of the Ancient Greeks. In particular, it was the Greeks who advanced the concept of the *Golden Mean*, and of the classical ideal of "Nothing to excess, but rather, moderation in all things."

Tracing back through the historical records, we repeatedly find the name of one Ancient Greek, especially, who is famous for this philosophy. His name is *Hippocrates* (hih-**PAHK**-rah-**tees**). Hippocrates was a physician, teacher, and author who lived in Greece from about 460–370 B.C. Hippocrates is widely considered the *Father of Modern Medicine*, because he broke with ancient superstitions and saw human health and disease in more rational, cause-and-effect terms.

Before his birth, earlier Greek philosophers interpreted the Cosmos as being composed of four basic *universal elements*. These were *air, fire, earth*, and *water*. To Hippocrates (and later, Aristotle), this balance in the large, external macrocosm (universe) was also reflected as a balance within the much smaller, internal microcosm (human body).

Specifically, Hippocrates believed in a *balance of the four body humors* (**HYOO**-mers) or "fluids." This balance was formally called *the Humoral* (**hyoo-MOR**-al) *Doctrine*. The four body humors were the four primary types of fluid the Ancient Greeks thought made up the body. (Consult Figure 2.1.)

The first body humor, *blood*, was hot and moist, like the moist universal element, *air*. The second body humor, *yellow bile* from the liver, also seemed hot, but was more dry than blood (due to its lower water content). Thus, the Greeks equated yellow bile with the universal element, *fire*. The third body humor, *black bile*, was mistakenly thought to come from the *spleen*. (This error may have arisen from the frequent clinical observation of black, blood-filled vomit or stools in patients suffering internal bleeding.) Since black bile was considered rather cold and dry, it was tied to the universal element, *earth*.

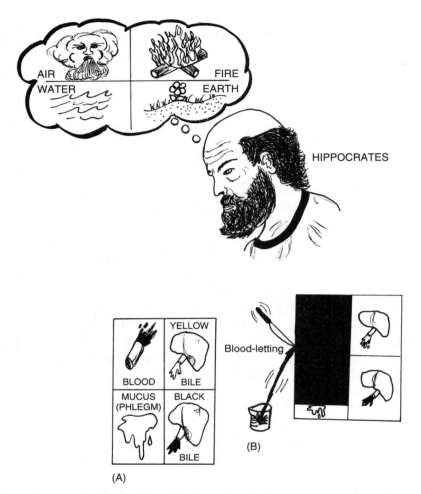

Fig. 2.1 Hippocrates and the Humoral Doctrine. (A) A balance of the four body humors (health). (B) Relieving a humoral imbalance by blood-letting.

Finally, the fourth body humor was *mucus* (**MYEW**-kus) or *phlegm* (**FLEM**). Like the clear, sticky discharge from a runny nose, the mucus (phlegm) was classified as cold and wet. Hence it was connected to the universal element, *water*. (The early Greeks mistakenly thought that mucus came from the brain, since the brain lies immediately superior and posterior to the nose!)

Hippocrates believed these four body humors were kept in a rough state of balance or equilibrium within the healthy patient (Figure 2.1, A). When a person became ill, Hippocrates reasoned, one of the four body humors (such as the blood) was present in a much greater amount than the other three. (Observe Figure 2.1, B.) A logical mode of treatment, therefore, was *blood-letting*. By cutting a vessel and letting out the excess blood, the Greek healers reasoned, the four body humors could be brought back into a rough equilibrium or balance, and a state of health be re-achieved.

Claude Bernard and His Dogs: A Preview of Homeostasis

The Golden Greek Ideal of the Humoral Doctrine (balance among the four body humors) did, when you really think about it, involve an equilibrium among body fluids, rather than solid structures.

Almost 2,000 years later, on the Continent of Europe, a Frenchman named *Claude Bernard* (ber-**NAR**) made his own experimental follow-up to Hippocrates and the Humoral Doctrine. Bernard was a vivisectionist, like Claudius Galen (Chapter 1) long before him. As an *experimental physiologist* (fiz-ee-**AHL**-uh-**jist**), Bernard, of course, badly needed to observe the functioning of living organisms. But to the great horror of his rich wife, who heard them howling, Bernard strapped dogs down onto lab tables in the basement of his private lab, and operated upon them while they were still alive! Even worse, Bernard did not give them any *anesthesia* (**an**-es-**THEE**-zha) to "remove" (*an-*) their pain "sensations" (*esthes*)! (Examine Figure 2.2.)

One of the main points of interest for Bernard was the physiology of the digestive system. It was during surgery of the open *abdominal* (ahb-**DAHM**-ih-nal) *cavity* of one of his howling dogs that Bernard made an important discovery. He found important information about the natural production of glucose – a "sweet" (*gluc*) "carbohydrate" (*-ose*), that is, a *sugar*. Glucose is the major sweet carbohydrate (sugar) used for fuel by the body's cells. Bernard collected the blood draining from the dog's liver through the *hepatic* (heh-**PAT**-ik) *veins*, which run below it. Now, Bernard had fed the dog only meat, which is mostly composed of protein. To his surprise, however, the blood draining from the dog's liver was rich in glucose! Since the dog hadn't eaten any carbohydrates, Bernard correctly reasoned, the liver must have *synthesized* (**SIN**-thuh-**sized**) the glucose itself, by putting the various parts of the molecule "together" (*syn-*).

This important observation provided the first strong evidence of natural chemical synthesis – the making of new chemicals within the body. The technical name for this particular type of chemical synthesis is *gluconeo-*

Cell 1

Cell 2

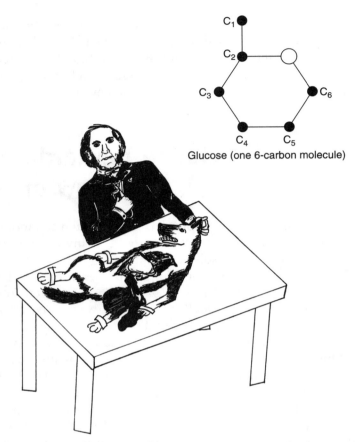

Glucose (one 6-carbon molecule)

Fig. 2.2   Claude Bernard finds glucose in a dog's bloodstream.

genesis (**gloo**-koh-**knee**-oh-**JEN**-uh-sis). Gluconeogenesis is literally the "production of" (*genesis*) "new" (*neo*) "glucose." This natural chemical synthesis of glucose by cells in the liver uses sources other than carbohydrates. [**Study suggestion:** Read back through the preceding section carefully. Now, ask yourself, "What was probably the type of chemical used by the dog's liver cells to help it synthesize the new glucose molecules Bernard found in its bloodstream?"]

MILIEU INTÉRIEUR: FRENCH FOR THE "INTERNAL ENVIRONMENT"

Beyond his specific observations and discoveries, Claude Bernard really gained a place for himself in history by stating an important general concept.

This concept is that of the *milieu* (**MEEL**-yoo) *intérieur* (an-**TARE**-ee-er), which is French for "internal environment."

In this book, we have considered the internal environment as essentially the same thing as Human Bodyspace. We have defined it in a very broad sense, as including everything in the body that lies deep to the surface of the skin.

Claude Bernard, however, had a more narrow interpretation of the internal environment (*milieu intérieur*). For him, it primarily consisted of the inner body fluids (including the blood), which provide a bathing medium for the cells. This bathing fluid includes important *nutrients* (**NEW**-tree-unts) or "nourishing substances," such as oxygen and glucose. The cells receive these nutrients, and utilize them for their physiology and metabolism. The *waste products* of metabolism, such as *carbon dioxide* (die-**OX**-eyed) and *acid*, are released from the cells, and pass into the internal environment.

Cell 3

To properly understand Bernard's concept of the internal environment, we need to take a closer look at the fluid both inside and outside of our cells. This will require some new terms. The prefix *extra-* means "outside of" (something). And the word *cellular* (**SELL**-you-lar) refers to "little cells." We can put this information together and create a single new term, *extracellular* (**eks**-trah-**SELL**-you-lar). The *extracellular fluid*, therefore, is the fluid that literally lies "outside the little cells." (See Figure 2.3.)

Since all of our cells lie deep within the body, the extracellular fluid surrounding the cells acts as Bernard's "internal environment." Even though it is located *outside* the cells, the extracellular fluid still lies *internal* or deep to the skin surface. Hence, this fluid is both internal (to the skin surface) and external (to the individual body cells), at the same time! Summarizing these ideas, we can write the following word equation:

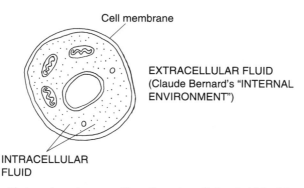

Cell membrane

EXTRACELLULAR FLUID
(Claude Bernard's "INTERNAL
ENVIRONMENT")

INTRACELLULAR
FLUID

Fig. 2.3 The "internal environment" as the extracellular fluid bathing our cells.

Cell 1

EXTRACELLULAR FLUID	=	Fluid lying outside and around the body cells, but still deep to skin surface	=	CLAUDE BERNARD'S "INTERNAL ENVIRONMENT"

"What about the fluid lying *inside* our body cells?" you may well be asking at this point. Taking the same approach to word-building, we create a new term, *intracellular* (**in**-trah-**SELL**-you-lar). The *intracellular fluid*, consequently, is the fluid present "within or inside" (*intra-*) our "little cells" (*cellul*).

Cell 2

Because both nutrients and waste products frequently pass into and out of the cell, many of the components of the intracellular fluid and extracellular fluid mix with one another. "It is the fixity of the internal environment that is the condition of free and independent life," Claude Bernard maintained. Because the components of the extracellular fluid (internal environment) are relatively stable, the physiology within our cells is likewise relatively stable. As Bernard discovered, for example, the concentration of glucose within the bloodstream (and other portions of the body's extracellular fluid) remains relatively constant over time. Hence, the body cells are assured of a fairly steady supply of glucose energy for their continued long-term survival.

Tissue 1

Walter B. Cannon: *The Wisdom of the Body* and *Homeostasis*

Claude Bernard (1813–1878) did most of his work in the mid-1800s. Another important physiologist, *Walter B. Cannon*, was born in 1871. Thus, Cannon was only seven years old when Claude Bernard died.

Molecule 1

Around 1915 Cannon began work on the body's response to severe stress. It was Cannon who showed how the *hormone* called *adrenaline* (ah-**DREN**-ah-lin) or *epinephrine* (**ep**-ih-**NEF**-rin) was released into the bloodstream during the so-called *"fight-or-flight" response*. By the "fight-or-flight" response, we mean the reaction of a human or animal to severe stress – and perhaps, life-threatening danger! So the frightened deer or rabbit being chased by a pack of howling wolves must either run away ("take flight") or just stand there and "put up a fight"!

Certainly, Cannon was vividly impressed by the linkage of strong emotions, such as those of rage and fear in the "fight-or-flight" response, to the increased release of adrenaline (epinephrine) into the human bloodstream.

Organ 1

And epinephrine, once released, was found to powerfully stimulate dramatic increases in heart rate, blood pressure, and various other body functions.

THE FIRST DEFINITION OF HOMEOSTASIS IS BORN

In 1926, Walter Cannon generalized his observations in an important book, called *The Wisdom of the Body*. It was in this book and associated research papers that Cannon showed the great "wisdom" to create the term *homeostasis*. His observation that the strong emotions and the hormone epinephrine greatly *disrupted* normal, fairly constant and stable, patterns of body functioning, logically led to the concept of homeostasis within Cannon's mind.

Walter Cannon defined homeostasis as "a condition which may vary, but which is relatively constant."

HOMEOSTASIS (Walter   =   A *RELATIVELY CONSTANT*
Cannon's original definition)   *CONDITION* WITHIN THE BODY'S
INTERNAL ENVIRONMENT

Homeostasis and Modern Control System Theory

Since the time of Claude Bernard and Walter Cannon, the techniques and instruments for measuring various aspects of the body have been greatly refined. Hence, we now speak of specific anatomical and physiological *parameters* (par-**AM**-uh-**ters**) within the body's internal environment. (A parameter, in general, is an aspect of some thing – such as oral body temperature – that has been measured and expressed as a certain number of units.) And the internal environment concept can be greatly broadened to include not only the extracellular fluid bathing our cells, but all of Human Bodyspace lying deep to the surface of the skin.

Recall (Chapter 1) that we record body parameters and see whether their current values lie within their normal range – the distance between their lower normal limit and their upper normal limit. Let us repeat and adapt from Figure 1.9, which showed the example of oral body temperature recorded in degrees Fahrenheit (Figure 2.4). We can say that a mouth temperature between 97.6 degrees F and 99.6 degrees F represents *normothermia* (**nor**-moh-**THERM**-ee-ah). This literally means "a condition of" (-*ia*) "normal" (*normo-*) body "heat" (*therm*) or temperature. Oral body temperature thus lies within its "normal" (*normo-*) range.

Organism 1

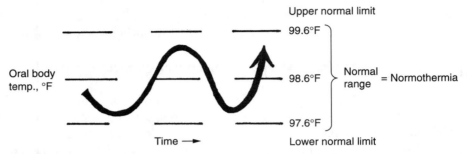

Fig. 2.4 Normothermia and homeostasis of oral body temperature.

Since the physiological parameter of oral body temperature in degrees F varies or changes over time, it cannot be described as constant. However, since the temperature does rise and fall in a roughly S-shaped pattern over time, staying within its normal range, we can legitimately say that it is *relatively* constant.

Hence, the oral body temperature recording displayed in Figure 2.4 is a case of homeostasis. A modern definition of homeostasis is thus:

HOMEOSTASIS = RELATIVE CONSTANCY OF A
(a modern definition) PARTICULAR ANATOMICAL OR
PHYSIOLOGICAL PARAMETER WITHIN
THE BODY'S INTERNAL
ENVIRONMENT

Various mechanisms, in both human bodies and machines, help guide and control their important structural and functional parameters, such that the parameters do, indeed, stay within their normal ranges. Logical explanations for this controlled, orderly behavior form a large part of what is often called *modern control system theory*. Comparisons between human nervous systems and mechanical control systems "piloted or steered" by computers are made in the science called *cybernetics* (*sigh*-ber-*NET*-iks). Cybernetics was originally defined as "the science of communication and control in man and machine." In our world, *mechanical* control systems usually involve various parameters that are monitored and regulated by computers. Hence by emphasizing *body* parameters we are engaging in what might be called *Compu-Think* – "*compu*ter-like modes or ways of human *think*ing."

The Set-Point Theory

Consider one type of mechanism, a *pendulum* (**PEN**-juh-lum) that is "hanging" down from a clock. After the pendulum has swung back and forth, tracing an arc, we will place the clock onto its side (Figure 2.5, A).

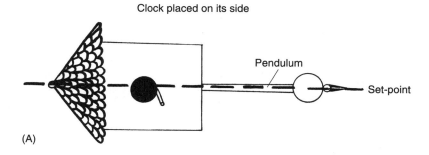

Clock placed on its side

(A)

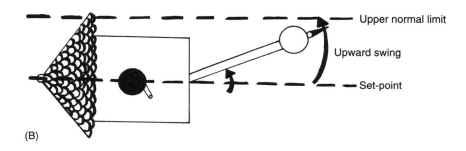

(B)

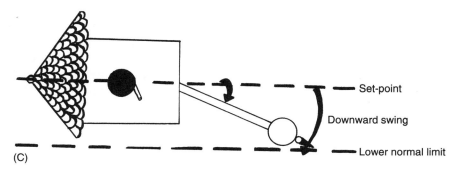

(C)

Fig. 2.5   The swinging clock pendulum and its set-point. (A) The set-point of a clock pendulum. (B) Pendulum swings toward its upper normal limit. (C) Pendulum swings toward its lower normal limit. (D) The normal range of pendulum swinging. (E) The S-shaped pattern of homeostasis (resembles the swinging of a pendulum).

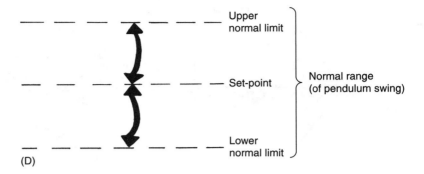

Fig. 2.5   (continued)

The *set-point* is the place or *point* in space where the pendulum and its weighted end were initially *set*. Over time, the pendulum swings upward, tracing an arc toward its upper normal limit of travel (Figure 2.5, B). The force of gravity then makes it swing back down, toward its set-point, before it traces another arc toward its lower normal limit of travel (Figure 2.5, C). This back-and-forth swinging, alternately above, then below, the set-point, creates an arched pattern around the set-point (Figure 2.5, D). This arched pattern somewhat resembles the S-shaped pattern we have proposed for homeostasis (Figure 2.5, E). In each case, there is movement up and down, but, long term, it always tends to center around the set-point.

In terms of a body parameter, the set-point is something different. The set-point is the long-term average value of a body parameter; that is, it is the *point* at which the parameter seems to be *set*. Take, for instance, the parameter of oral body temperature. Its set-point is about 98.6 degrees F. If a person didn't know her oral body temperature at any given time, her best guess would be 98.6 degrees F, because, on average, she would be right on this set-point value!

Stressors or Stimuli: Disturbers of Body Parameters

A person with a probing mind might now question, "Well, if oral body temperature is actually *set* at a point or value of about 98.6 degrees F, then why doesn't it just *stay* there? Why don't we have an *absolute constancy* of oral body temperature over time, with no change in its value whatsoever?" The answer is that numerous *stressors* (**STRESS**-ors) just won't *let* the parameter stay at its set-point value for very long!

A stressor is a change in the internal or external environment that disturbs a body parameter from its set-point level. For oral body temperature, there are *heat stressors* that tend to push the value of oral body temperature above its set-point level. Drinking a cup of hot coffee, taking a warm bath, or vigorously exercising, for instance, could all act as heat stressors that push the body temperature up toward its upper normal limit of 99.6 degrees F. (Consult Figure 2.6, A.)

Conversely, drinking a glass of cold water, taking a cool shower, or resting on your back stripped down to your jockey shorts, could all act as *cold stressors*. These stressors would tend to push the value of oral body temperature below its set-point level, and toward its lower normal limit of 97.6 degrees F. (Examine Figure 2.6, B.)

Therefore, since stressors are present everywhere, all the time, absolute constancy of a particular body parameter is an unattainable goal. The best you can reasonably hope for is that a relative constancy of the body parameter – homeostasis – can somehow be maintained.

Certain stressors are also *stimuli* (**STIM**-you-**lie**). A stimulus or "goad" (mentioned in Chapter 1) is a *detectable* change in the body's internal or external environment. A cold stressor, such as having a pail of icy water dumped over your head, disturbs relative constancy of oral body temperature, but it acts as a stimulus as well. How do we know this? – We can *feel* or *sense* it! [**Study suggestion:** Suppose there was a 500-pound block of radioactive material that someone hid, without your knowledge, under your desk at work or at school. Would you still come and sit there every day? Would this radioactive material be a significant stressor? Would it also act as a stimulus? – Why, or why not?]

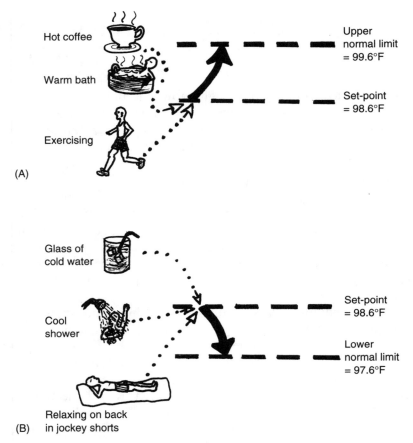

Fig. 2.6 Contrasting effects of heat stressors versus cold stressors upon oral body temperature. (A) Heat stressors that raise body temperature above its set-point level. (B) Cold stressors that lower body temperature below its set-point level.

Feedback Systems or Cycles

Let us return to a consideration of an ordinary clock pendulum. Interestingly enough, temperature stressors can also "disturb" a clock pendulum! When the room temperature gets hot, the rod in a clock pendulum expands, and it swings more slowly. Conversely, when the room gets cold, the rod in the pendulum contracts, and it swings more rapidly. But no matter how fast or slowly the pendulum swings, it has a particular normal range of travel, swinging first to an upper normal limit at one end of its arc, and another at the opposite end of its arc. The regulation and *control* of the distance of swinging of such a clock pendulum from its set-point position, therefore, is a relatively simple matter.

THE GENERAL OPERATION AND COMPONENTS OF FEEDBACK CYCLES

The regulation and control of various body parameters, such as oral body temperature in degrees Fahrenheit, however, is much more complex and involved! This regulation is accomplished by the action of what are formally called *feedback systems* or *feedback cycles*. A feedback system or cycle is a system whose output or response curves or "feeds" back upon the start (the stimulus or stressor), thereby influencing whether the system repeats itself or not. A general plan for a feedback system or cycle is traced in Figure 2.7, A.

There are two main types of feedback systems or cycles. The basic operation of a *negative feedback system* is shown in Figure 2.7 (B), while that for a *positive feedback system* is depicted in Figure 2.7 (C).

A NEGATIVE FEEDBACK OR CONTROL SYSTEM

A negative feedback system is a system whose output or response curves or "feeds" back upon the start in a *negative* manner, by removing or correcting the conditions of the original stimulus. (This "negative" or "removing" effect upon the original stimulus is symbolized by the minus sign, and by a red traffic light in Figure 2.7, B, which essentially tells the driver of a car, "Stop, Baby! Stop!")

A group of sensory receptors originally sense or detect the stimulus (Figure 2.8). The sensory receptors, after they are excited, send information over an *afferent* (**A**-fer-**ent**) or *sensory pathway*. An afferent or sensory pathway is a pathway that "carries" (*fer*) information about the stimulus from the sensory receptors, and "toward" (*af-*) a *control center*. A control center is a collection of anatomical and physiological components that establishes and maintains the set-point for a particular body parameter. A signal from the control center is then sent over an *efferent* (**EE**-fer-**ent**) or *motor pathway*. An efferent or motor pathway is a pathway that "carries" (*fer*) motor or "movement"-related information "away" (*ef-*) from the control center, and out toward body effectors.

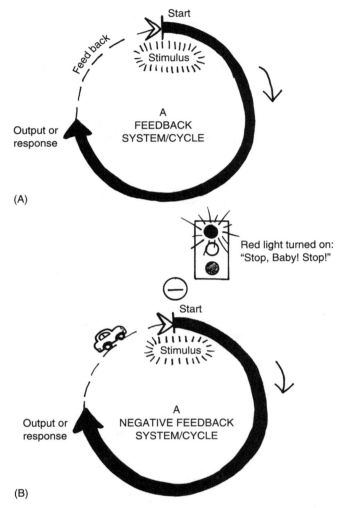

Fig. 2.7 Feedback systems or cycles. (A) General operation of any feedback system/cycle. (B) A negative feedback system/cycle. (C) A positive feedback system/cycle.

When the body effectors are stimulated, they carry out some kind of response. Finally, the response of the negative feedback system curves or feeds back to remove or correct the conditions of the original stimulus. Because the negative feedback system generally removes or corrects the stimulus, the body parameter involved is usually controlled enough to stay within its normal range. Thus, another name for a negative feedback system is a negative feedback *control* system.

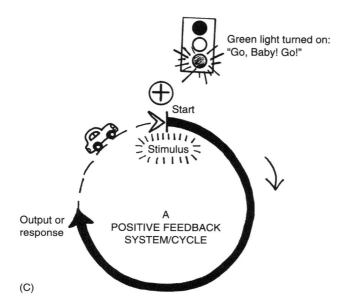

Green light turned on:
"Go, Baby! Go!"

Start

Stimulus

A
POSITIVE FEEDBACK
SYSTEM/CYCLE

Output or
response

(C)

Way back in Chapter 1, we introduced some of the components of a negative feedback control system that were involved in *thermoregulation* (**ther**-moh-reg-you-**LAY**-shun). This is literally the "regulation" of body "heat" (*therm*) or temperature. Figure 2.9 will now do this in much greater detail.

Assume that you decide to take a run. Very soon, your oral body temperature starts to rise towards its upper normal limit. This is the original

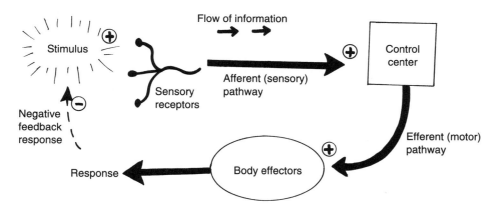

Fig. 2.8 The general components of a negative feedback control system.

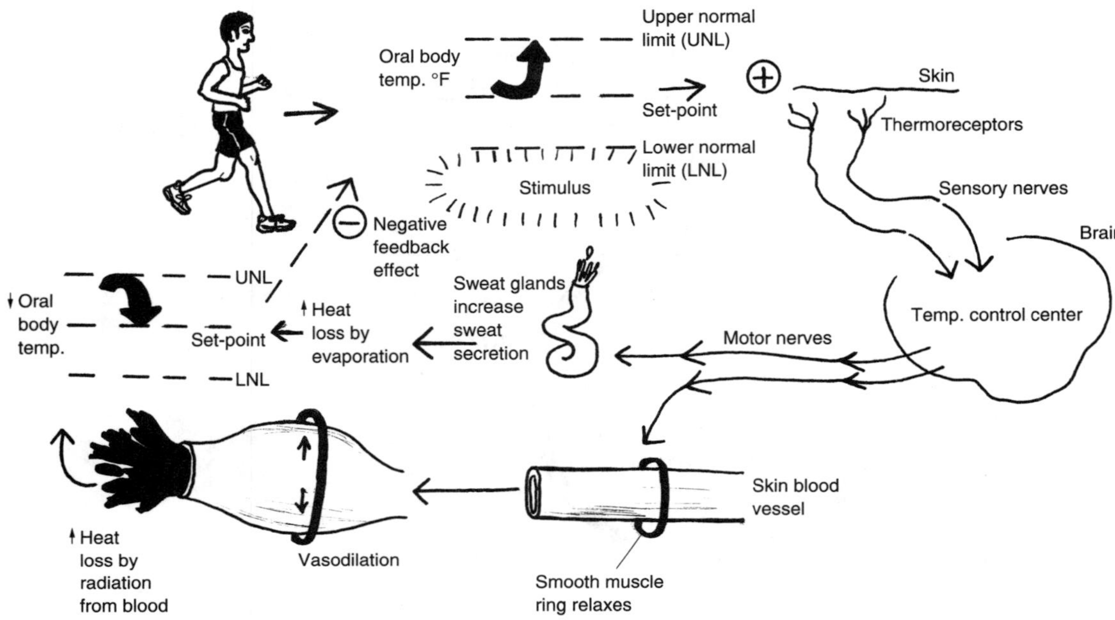

Fig. 2.9 Operation of a negative feedback control system that "corrects" a rise in body temperature.

stimulus or stressor. This rise in temperature stimulates skin thermorecep-tors, free nerve endings that are sensitive to heat. Afferent (sensory) nerve fibers carry signals from the skin thermoreceptors over *sensory nerves*, and to the *temperature control center* in the brain.

It is the temperature control center in the brain that establishes and main-tains the set-point for oral body temperature at about 98.6 degrees F. Once activated by the sensory nerves, the temperature control center sends a cor-rection signal out over *motor nerves*. These motor nerves serve as the efferent pathway, which carries the correction signal back toward the skin.

Two types of effectors in the skin make helpful responses. *Sweat glands* increase their rate of sweat secretion. The cool sweat evaporates from the skin surface, taking some of the body heat along with it. A second type of effector is the ring of *smooth muscle* surrounding the walls of skin blood vessels. When this ring of smooth muscle relaxes, it is like the loosening of a noose around someone's neck. The loosened noose allows *vasodilation* (**vase**-oh-die-**LAY**-shun) to occur. This is literally the "process of" (-*tion*) blood "vessel" (*vas*) "widening" (*dilat*).

The wider vessels allow more hot blood to leave the deep inner core of the body, and shunt out toward the skin. Here much more body heat is lost by *radiation* of heat waves from the blood near the skin surface, out into the surrounding air.

The final response of this heat-correcting system is a negative feedback loop. The evaporation of sweat plus the vasodilation of skin blood vessels both combine to drop the oral body temperature back toward its set-point level of 98.6 degrees F. Since this feedback is "negative," it prevents or discourages another round of sweating and vasodilation, because it removes or corrects the condition of the original stimulus.

Important to note is that a similar negative feedback control system operates at the lower end of the temperature scale, when oral body temperature drops toward its lower normal limit. [**Study suggestion:** For a stimulus or stressor of this type, take a piece of paper and try to write out the exact *opposite* sequence of events that occurred to a rise in oral body temperature toward its upper normal limit. Check your thinking with an educated friend.]

In summary, a negative feedback control system is "negative" in the sense that it is *error-minimizing*. The negative feedback system doesn't really "care" whether a body parameter is rising up toward its upper normal limit, or falling toward its lower normal limit. Both of these changes from the set-point level represent *errors* in the parameter, usually due to the disturbing effects of various stressors. The negative feedback system controls the parameter by minimizing or reducing the magnitude of either upward or downward errors. In this way, it provides a critical mechanism for maintaining homeostasis (relative constancy) of the body parameter involved.

A NEGATIVE FEEDBACK CONTROL SYSTEM	=	An "Error-Minimizing" or "Error-Reducing" Correcting Loop That Helps Maintain Homeostasis of Body Parameters, Even in the Face of Disturbing Stressors!

A POSITIVE FEEDBACK OR "OUT-OF-CONTROL" SYSTEM

A positive feedback system (review Figure 2.7, C) is modeled using a plus (+) sign at the tip of the feedback arrow, as it curves back upon the stimulus or stressor. And our little car on the racetrack encounters a green light, telling the driver, "Go, Baby! Go!"

A positive feedback system or cycle is a system whose output or response curves or "feeds" back upon the start in a *positive* manner, by magnifying or speeding up the conditions of the original stimulus. Let us now examine how a positive feedback system works. As for negative feedback, we will use as our example changes in oral body temperature from its set-point level.

Hyperthermia: body temperature rises "out of control" at the high end

Again, assume we go out and start running. Very soon, our oral body temperature starts to rise toward its upper normal limit. But as the temperature approaches this ceiling, a positive feedback (rather than negative feedback) process kicks into operation. (View Figure 2.10, A.) This means that the increase in body temperature above its set-point level just keeps rising and rising, until a *hyper-* state, one that is "above normal or excessive," is reached. The resulting condition is technically called *hyperthermia* (**high-per-THER-me-uh**) – "a condition of" (*-ia*) "above normal or excessive" (*hyper-*) body "heat" (*therm*) or temperature. In short, the positive feedback system just keeps on boosting the temperature higher and higher, far beyond 99.6 degrees F. Obviously, this dangerous and unstable condition of hyperthermia cannot keep going on forever! Sooner or later, a dramatic climax is reached. The internal body temperature may cause the feverish person to pass out, or even fall into a coma and die of brain damage!

Organism 1

Hypothermia: body temperature falls "out of control" at the low end

"Since hyperthermia is so potentially dangerous," you may now be asking yourself, "shouldn't we be hoping for the exactly *opposite* body state?" Speaking scientifically, this opposite body state is called *hypothermia* (**high-poh-THER-me-uh**). The prefix *hypo-* means "below normal or deficient." [**Study suggestion:** Following directly from the procedure used to literally translate hyperthermia, try to come up with an exact translation for hypothermia.]

Imagine that it is wintertime. The streets are covered with ice and snow, and they are very slippery. Your car slides off an isolated country road, and into a ditch. It is after midnight, and you begin to walk out into the darkness, seeking help. You are then at great risk for developing hypothermia, aren't you! As your body just keeps getting colder, your oral temperature declines toward the lower normal limit of 97.6 degrees F. (Study Figure 2.10, B.)

When it hits 97.6 degrees F, a positive feedback system is set into operation. Therefore, the colder the body gets, the more slowly you walk down the

Fig. 2.10   Positive feedback effects upon body temperature: creation of hyperthermia (A) versus hypothermia (B).

road. And the more slowly you walk down the road, the colder your body gets. The temperature drops below the lower normal limit, and just keeps falling and falling! Of course, a dramatic climax will eventually be reached! – What happens? "I feel so tired," you mutter to yourself, "Maybe I just need to lie down and rest for a while!" *Bad idea!* Severe hypothermia sets in, and you go into a coma and die! (They find your stiff, frozen body buried in the snow, the next morning.)

In summary, a positive feedback system is "positive" in the sense that it is *error-maximizing*. The positive feedback system doesn't really "care" whether a body parameter is rising toward its upper normal limit, or falling toward its lower normal limit. Both of these changes represent *errors* in the body parameter – upward or downward deviations or changes from the set-point level. The positive feedback system just keeps making these upward or downward errors bigger and bigger! This is why the positive feedback system is alternatively called a *vicious circle* or *vicious cycle*. Like a giant snowball rolling downhill, the bigger the snowball (error in a body parameter) gets, the faster it rolls, and the faster the snowball rolls, the bigger it gets! There is just no stopping this giant snowball, is there? It will keep getting bigger and bigger (the parameter going ever farther into a hyper- or hypo- state) until a dramatic climax is reached. (Maybe the giant snowball finally crashes into the side of a building!)

Positive feedback systems (vicious cycles or vicious circles) are definitely *not* control systems! Rather, they are dramatically *"out-of-control"* systems! Hence, they tend to greatly disrupt homeostasis of body parameters, rather than maintain them.

A POSITIVE = An "Error-Maximizing" or "Error-Increasing"
FEEDBACK SYSTEM Vicious Cycle That Usually Disrupts
Homeostasis of Body Parameters, Pushing
Them "Out of Control"

Feedback Systems and Human Health

Organism 2

From the above discussion, we can conclude that the type of feedback system involved – positive versus negative – can greatly influence human health and well-being. Technically speaking, a state of *clinical* (**KLIN**-ih-kal) *health* is one where the body is basically healthy, so that there is no legitimate reason to visit a "clinic" or physician's office for treatment. A state of Biological Order of the important body parameters is being maintained. This is generally reflected in the occurrence of homeostasis (relative constancy) of these parameters. And as we have seen from the time of the Ancient Greeks (such as Hippocrates), human health is usually seen as a balance. Using modern control system theory, we could also say that the negative feedback control systems are the ones that often maintain this balance or relative constancy of the various body parameters. This means that the body parameters stay within their *normo-* ("normal") *range or normo zone* over time.

For specific examples of clinical health and homeostasis, let us consider both oral body temperature in degrees Fahrenheit, as well as blood glucose concentration, measured in milligrams of glucose per 100 ml (milliliters) of blood.

Recall from earlier in this chapter that normothermia is a condition of normal body heat or temperature. This normothermia essentially involves a state of the entire organism, since the temperature measured in the mouth generally represents the temperature in the entire human body. When this normothermia is maintained over time, then we can say that this high degree of Biological Order is an instance of homeostasis or relative constancy, maintained by negative feedback mechanisms. And the oral body temperature can be legitimately described as clinically healthy.

Remember the work of Claude Bernard on the relative constancy of the body's "internal environment" (for him, the extracellular fluid bathing our cells). This included a careful measurement of blood glucose concentration. This anatomical parameter is that of a particular type of molecule – glucose – and its concentration within the bloodstream. Just as we used normothermia to describe oral body temperature within its normal range, we can coin the term *normoglycemia* (**nor**-moh-gleye-**SEE**-me-ah). This term indicates a "blood condition of" (*-emia*) "glucose" (*glyc*) within its "normal" (*normo-*) range. Under fasting conditions, the normal range for blood glucose concentration is from a lower normal limit of 70 mg/100ml blood, to an upper normal limit of about 110 mg/100 ml blood. When the amount of upward and downward change from the set-point value of 90 mg/100 ml blood stays relatively "balanced" between these limits over time, then a state of homeostasis of blood glucose concentration is said to exist.

Tissue 1

Both the *normothermic* (**nor**-moh-**THERM**-ik) and *normoglycemic* (**nor**-moh-gleye-**SEE**-mik) conditions represent states of clinical health of these critically important body parameters. Again and again in this book, we will see that other body parameters lying in their *normo-* ("normal") ranges likewise represent states of homeostasis and clinical health.

POSITIVE FEEDBACK, MORBIDITY, AND MORTALITY

Positive feedback, as we have seen, maximizes or increases the amount of upward or downward "error" from the set-point level of a particular body parameter. Positive feedback at the upper normal limit commonly results in a *hyper-* ("above normal or excessive") state of the body parameter. In the case of body temperature, this hyper- state would be called hyperthermia. (Review Figure 2.10, A, if desired.) And in the case of blood glucose concentration, this above-normal or excessive state is called *hyperglycemia* (**high**-per-gleye-**SEE**-me-ah).

Tissue 1

At the opposite end of the normal range, positive feedback at the lower normal limit creates a *hypo-* ("below normal or deficient") state of the body parameter. For oral body temperature, of course, this hypo- state is called

Tissue 2

hypothermia. (Review Figure 2.10, B.) And for deficient or below-normal blood glucose concentration, the scientific term is *hypoglycemia* (**high**-poh-gleye-**SEE**-me-ah). Either of these conditions results in a loss or disruption of homeostasis in the parameters involved.

Morbidity and mortality

Either a hyper- state or a hypo- state typically involves "diseased" (pathological) conditions of body structures and functions. This means pathological anatomy and pathophysiology. Such deficient or excessive pathological states are lumped into a general category called *morbidity* (**more-BID**-ih-tee) – "conditions of" (*-ity*) "illness" (*morbid*). If these *morbid* (**MOR**-bid) or "ill" states become very severe, they may result in *mortality* (**more-TAL**-ih-tee) – "a condition of" (*-ity*) "death" (*mortal*).

POSITIVE FEEDBACK LEADING TO "HELPFUL" CLIMAXES

Our discussion of positive feedback systems (vicious circles or vicious cycles of ever-increasing amounts of change) has generally been an unhappy one. From the important perspective of our clinical health, the dramatic climaxes and extreme hyper- or hypo- states reached by positive feedback mechanisms have been harmful. By disrupting the homeostasis of important body parameters, the positive feedback cycles often result in morbidity (illness) or mortality (death).

Certainly, illness and death are dramatic climaxes, aren't they? There are a few noteworthy examples, however, where positive feedback systems leading to dramatic climaxes are most helpful (rather than harmful) to human survival. These include: *sexual climax or orgasm, contractions of the womb leading to childbirth,* firing of *nerve impulses,* and *clotting of the blood* after vessel injury. We will discuss these specific examples in more detail in later chapters. But for now, ask yourself this important question about blood clotting: "I have suffered cuts in my skin many times during my life. How come all of these cuts (say, from injuries suffered more than a month ago) aren't still oozing blood today? What type of feedback process has made all of this blood-leaking shut off for good?"

Quiz

Refer to the text in this chapter if necessary. A good score is at least 8 correct answers out of these 10 questions. The answers are listed in the back of this book.

1. Widely acknowledged as the Father of Modern Medicine:
    (a) Socrates
    (b) Arnold Schwarzenegger
    (c) Hippocrates
    (d) Galen

2. The original rationale behind the medical practice of blood-letting was:
    (a) An excess of blood was upsetting the balance of body humors
    (b) The idea that blood was a natural body poison
    (c) Fear that excessive blood glucose would lead to diabetes
    (d) A common consensus that blood was an unimportant body fluid

3. Baby Heinie swallows a bucket of tomatoes and vinegar, which is high in acetic acid. He then suffers a bad case of acid indigestion. The acetic acid thus seems to be acting as a:
    (a) Control system
    (b) Stressor
    (c) Set-point
    (d) Normal range of a body parameter

4. The concept of the *milieu intérieur* developed by Claude Bernard mainly refers to the:
    (a) Intracellular fluid and its dissolved particles
    (b) Four body humors of Ancient Greece
    (c) Secretions of the kidney into the urine
    (d) Extracellular fluid surrounding body cells

5. Walter Cannon's original definition of homeostasis:
    (a) The "fight-or-flight" response to severe stress
    (b) An automatic secretion of adrenaline (epinephrine) into the bloodstream whenever body parameters become disturbed
    (c) A relatively constant condition within the body's internal environment

(d) A state of absolute constancy of many of the body parameters over time

6. "Living organisms are machines" would best summarize:
   (a) The attitude of the early Chinese thinkers
   (b) The perspective of cybernetics and modern control system theory
   (c) A false rejection of the principles of the Humoral Doctrine
   (d) The "fight-or-flight" response

7. A relative constancy of a particular anatomical or physiological parameter within the body's internal environment:
   (a) Current definition of homeostasis
   (b) Outmoded superstitions about "ghosts" within the body
   (c) Description of the hyper- zone
   (d) A condition of illness

8. The long-term average value of a body parameter:
   (a) Normal range
   (b) Set-point
   (c) Upper normal limit
   (d) Chronic chaos

9. A negative feedback system generally operates to:
   (a) Restrain, inhibit, or minimize further change in a body parameter, in some particular direction
   (b) Create "negative" conditions within the internal environment that are harmful to survival
   (c) Stimulate, magnify, or accelerate the distance between the current value of a body parameter and its set-point value
   (d) Halt any up-or-down movements whatsoever of a body parameter from its set-point level

10. Most likely involved in establishing trends toward morbidity or mortality:
    (a) Normal body structures and functions
    (b) States of pathophysiology and pathological anatomy
    (c) Positive feedback mechanisms that promote blood clotting after vessel injury
    (d) Negative feedback control systems operating at peak efficiency

Body-Level Grids for Chapter 2

Several key body facts were tagged with numbered icons in the page margins of this chapter. Write a brief summary of each of these key facts into a numbered cell or box within the appropriate *Body-Level Grid* that appears below.

Anatomy and *Biological Order* Fact Grids for Chapter 2:

A

CELL
Level

1	2

TISSUE
Level

1

Anatomy and Biological Disorder Fact Grids for Chapter 2:

TISSUE
Level

1	2

Physiology and Biological Order Fact Grids for Chapter 2:

MOLECULE
Level

1

CELL
Level

1	2

3

TISSUE
Level

1

ORGANISM
Level

1	2

Physiology and Biological Disorder Fact Grids for Chapter 2:

ORGAN
Level

1

ORGANISM
Level

1

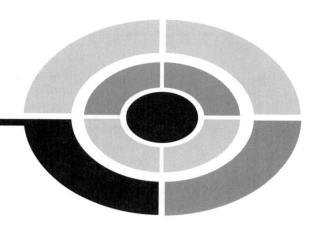

Test: Part 1

DO NOT REFER TO THE TEXT WHEN TAKING THIS TEST. A good score is at least 18 (out of 25 questions) correct. Answers are in the back of the book. It's best to have a friend check your score the first time, so you won't memorize the answers if you want to take the test again.

1. The word biology exactly translates to mean:
    (a) "Study of body functions"
    (b) "Study of life"
    (c) "Removal of living things"
    (d) "A cutting apart"
    (e) "Two kinds of living things"

2. Biological Disorder equals:
    (a) A break in all natural patterns
    (b) The establishment of normal ranges
    (c) The application of parameters to living things only
    (d) A disruption in living patterns
    (e) Birds who fly in a straight path through the sky

3. Body structures can be considered as _____ within a sentence:
   (a) Verbs
   (b) Adverbs
   (c) Adjectives
   (d) Periods
   (e) Nouns

4. Aristotle created the word physiology to originally mean:
   (a) "The study of Nature"
   (b) "Body structures and the study of body structures"
   (c) "A state of relative constancy"
   (d) "Body functions and the study of body functions"
   (e) "The process of dissection"

5. Blood pressure is measured and observed to remain between 90/60 millimeters of mercury and 140/90 millimeters of mercury, over a 24-hour period. We can logically conclude that:
   (a) No stressors were acting upon the blood pressure during this time
   (b) Homeostasis of blood pressure was being maintained
   (c) The blood pressure was absolutely constant over this period
   (d) Blood glucose concentration likewise remained within its normal range
   (e) The blood was far too hot to permit accurate pressure readings

6. The sole of the foot is rather broad and flat, allowing it to bear weight easily:
   (a) Law of Use and Disuse
   (b) Example of deadly force
   (c) Complementarity of body structure and function
   (d) An evolutionary error
   (e) Complained about by Aristotle

7. A stimulus is:
   (a) Always the same as a stressor
   (b) A detectable change in the body's internal or external environment
   (c) A specific kind of sensory receptor
   (d) Usually the last event in a negative feedback cycle
   (e) Latin for "goat" with "horns"

8. A level of body organization
   (a) Involves a particular size and complexity of body structures
   (b) Is defined in the book as the entire Pyramid of Life

(c) Starts with the cellular level

(d) Bears no relationship to the Law of Complementarity

(e) Just says something important about anatomy, not about physiology

9. Hyperthermia generally reflects:
   (a) Precise regulation of body temperature
   (b) A dramatic drop in blood glucose concentration
   (c) A sudden fall in oral body temperature below its normal range
   (d) Proper maintenance of homeostasis
   (e) A loss of control of body temperature at its high end

10. A good example of a physiological parameter:
    (a) Femoral length in cm
    (b) Blood glucose concentration in mg glucose/100 ml blood
    (c) Density of nerve cells in the brain, in number of cells/cubic centimeter of brain tissue
    (d) Weight of the liver in ounces
    (e) Rate of urine flow from the right kidney, in cc/minute

11. A plain functional parameter:
    (a) Heart rate in beats/minute
    (b) Car speed
    (c) Heat of exhaust emitted from a dumptruck tailpipe, in degrees celsius
    (d) Lung capacity in liters
    (e) Bladder circumference in inches

12. The word anatomy exactly translates to mean:
    (a) "a process of cutting up or apart"
    (b) "a penny for your thoughts"
    (c) "inflammation of a sweetie"
    (d) "production of new glucose"
    (e) "presence of body structures (not functions)"

13. The Ancient Greek idea of the Golden Mean:
    (a) A reflection of severe Biological Disorder
    (b) The product of unbalanced body humors
    (c) Was never supported in later eras
    (d) Is quite similar to the modern notion of the set-point
    (e) Has no valid application to current biological thinking

14. The Humoral Doctrine:
    (a) Assumed a balance among blood, yellow bile, black bile, and phlegm
    (b) Believed that "The Devil is in the details!"
    (c) "Eat, drink, and be merry; for tomorrow we may die"
    (d) Had no connection to the universal element of water
    (e) Died out before the time of the American Revolution

15. The highest level of a normal body parameter:
    (a) Upper normal limit
    (b) Set-point value
    (c) Lower normal limit
    (d) Normal range
    (e) Subnormal range values

16. Claude Bernard used vivisection during his experiments because:
    (a) He was a notable animal rights activist
    (b) His rich wife refused to give him enough money to feed his lab animals
    (c) He enjoyed torturing living dogs
    (d) Physiological mechanisms can only be observed in living organisms
    (e) Not enough dog and pig cadavers were available

17. The production of new molecules of glucose from non-sugar sources:
    (a) Abdominocentesis
    (b) Gluconeogenesis
    (c) Glycogenolysis
    (d) Enzyme-splitting
    (e) Glucose uptake from the intestine

18. The *milieu intérieur*:
    (a) An old German fairytale
    (b) A French phrase for "internal environment"
    (c) First advanced by Walter B. Cannon
    (d) Intensely denounced by Hippocrates
    (e) The famous brainchild of Edward Jenner

19. Large amounts of adrenaline (epinephrine) released into the bloodstream:
    (a) Normal, resting body conditions
    (b) A slow response to sudden danger
    (c) The "fight-or-flight" response

(d) A quick return to hormonal equilibrium
(e) Triggers a dramatic drop in body temperature

20. Baby Heinie pulls his Big Sister's pigtails; Big Sister slaps Baby Heinie; Baby Heinie pulls Big Sister's pigtails even harder; and Big Sister slaps Baby Heinie with even more force:
(a) An imaginative story about negative feedback
(b) A crucial "fight-or-flight" survival mechanism in operation
(c) Normothermia of blood pressure is being maintained
(d) Homeostasis of relevant body parameters
(e) An entertaining real-life scenario of a positive feedback cycle

21. The falling of blood pressure into the range of *hypotension* (**high**-poh-**TEN**-shun):
(a) The occurrence of set-point adjustment
(b) The "survival of the fittest"
(c) A clinical example of the Humoral Doctrine
(d) Probable starting of positive feedback near the lower normal limit
(e) Negative feedback promoting blood pressure homeostasis

22. A clock pendulum swinging back and forth from its initial vertical position:
(a) Demonstration of the normal range concept for body parameters
(b) A crude error in cybernetics
(c) Rough model for the idea of morbidity
(d) The removal of all feedback processes
(e) A reflection of actual real-life vicious circles

23. The temperature control center:
(a) Carries sensory information over an afferent pathway
(b) Delivers a response from the body effectors in the skin
(c) Monitors and adjusts the current body temperature to match the set-point level
(d) Removes all inhibitory influences upon blood glucose homeostasis
(e) Is a hypothetical body structure that does not really exist

24. A distraught mother rushes into a hospital emergency room and screams at the top of her lungs while the nurses are trying to apply a Bandaid to a small scrape on her young daughter's knee:
(a) Good instance of a movement toward parameter equilibrium
(b) An upsetting illustration of emotions rocketing into a hyper- state
(c) Dumbing-down of medical therapy so that a lay person can understand it

(d) Return of a previously morbid condition back into the normo-range

(e) Relative constancy around the "fight-or-flight" set-point

25. A single blood sample (taken at 8 A.M.) shows a glucose concentration of 92 mg/100 ml blood. From this fact you can most appropriately conclude that:

(a) Homeostasis of blood glucose concentration is being maintained

(b) The tested patient is suffering from hypoglycemia

(c) A diabetic coma is about to happen

(d) Normoglycemia existed at 8 A.M.

(e) Clinical health of the patient is guaranteed

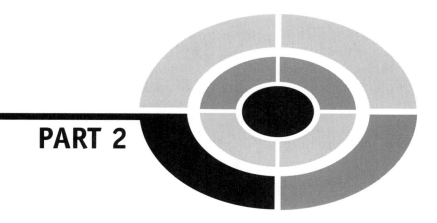

PART 2

Functions at the Chemical and Cellular Levels

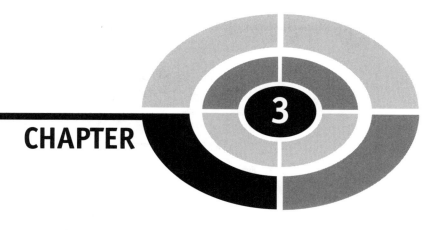

Chemical Functions: ''Hey! We've Got Dirt *Moving Around* in Our Body Basement!''

In Part 1 of *PHYSIOLOGY DEMYSTIFIED*, our aim was to present you with the underlying history and basic concepts behind the modern study of body functions. Chapter 2, Control of the Internal Environment, brought to the forefront the critical behind-the-scenes action of physiological control systems that help maintain homeostasis.

We will now, however, re-focus our vision upon the three bottom layers of the Great Body Pyramid. If you will remember, these were Levels I, II, and III, which, taken all together, make up the *Chemical Level* of Body Organization:

$$\begin{array}{c} \text{THE CHEMICAL} \\ \text{LEVEL} \end{array} = \begin{array}{c} \text{SUBATOMIC} \\ \text{PARTICLES} \end{array} + \text{ATOMS} + \text{MOLECULES}$$

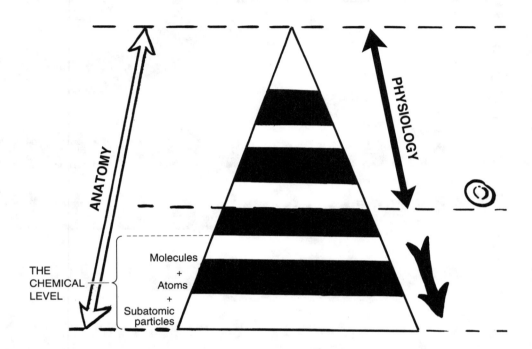

The Atom: An "Uncuttable" Particle with Its Own Anatomy

When the Ancient Greeks (such as Aristotle) talked about the internal micro-cosm or "tiny universe" within the human body (Chapter 1), how far *down* did they go? Not having any microscopes, of course, their ability to actually view and study objects at the micro- level was very limited! Nevertheless, their brilliant philosophy and logic allowed them to get quite close to describing the real identities of many extremely tiny body structures.

DEMOCRITUS DESCRIBES THE ATOM

Before Aristotle founded Natural History and began a thorough study of the human and animal body, another early Greek philosopher was wondering about the Ultimate Nature of All Things. About five hundred years B.C., *Democritus* (dih-**MAHK**-ruh-**tus**) first proposed the existence of atoms. He said that, "By convention sweet is sweet, by convention bitter is bitter, by convention hot is hot, by convention cold is cold, by convention color is color. But in truth there are atoms and the void."

What Democritus meant by this last statement was that all real objects in the world (including the structures of the human body) could be progressively broken down into smaller and smaller parts. But ultimately, these body structures consisted of only two things: "atoms and the void" (empty space around and between the atoms).

Birth of the Atomic Theory

We can give much of the credit to Democritus, then, for creating the basic foundations of *the Atomic* (uh-**TAHM**-ik) *Theory*.

THE ATOMIC THEORY AND HUMAN ANATOMY:

ALL STRUCTURES	=	ATOMS	+	EMPTY SPACE
		Therefore,		
ALL *BODY* STRUCTURES (*ANATOMY*)	=	*BODY* ATOMS	+	THE EMPTY SPACES BETWEEN THE *BODY* ATOMS

Atoms 1

Even now, about 2,000 years after Democritus, scientists still maintain that the basic principles of the Atomic Theory are true! Using today's advanced technology, scientists can actually *see* atoms with a special instrument called the *electron* (e-**LEK**-trahn) *microscope*. In 1803, a chemist named *John Dalton* actually proposed the modern Atomic Theory. Dalton, much like Democritus, believed that all of the chemical *elements* (fundamental types of matter) are composed of atoms, which are "not" (*a-*) "cuttable" (*tom*) into any smaller parts. He also believed that these atoms ("uncuttable" particles) were round in shape. Indeed, the *carbon* (C) atom, for example, which makes up the element carbon, does somewhat look like a fuzzy, round sphere (Figure 3.1, A).

(A)

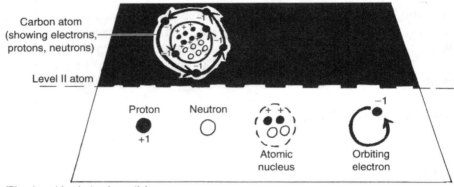

(B)   Level I subatomic particles

Fig. 3.1   The "dissectable" carbon atom and its subatomic particles. (A) The carbon atom as a ball of cotton candy or fuzzy sphere. (B) The subatomic particles present within a carbon atom.

SUBATOMIC PARTICLES WITHIN THE "UNCUTTABLE" ATOM

How fascinating it is to realize that

ANATOMY (BODY STRUCTURE) = ONE OR MORE BODY *ATOMS* AND THEIR *SMALLER PARTS* (Including the Empty Spaces Between These Parts)

As you may remember (Chapter 1), the very word anatomy literally translates from Ancient Greek to mean "the process of cutting (the body) up or apart." To be sure, anatomy has largely been studied throughout the centuries by means of dissection into smaller and smaller pieces. When we work down to the level of the atom, we find (from Democritus) that it was originally thought to be "not" (*a-*) "cuttable" (*tom*) or dissectable into any smaller parts.

With the coming of the twentieth century, however, scientists were finally able to split the supposedly uncuttable atom. Therefore the body atoms, like all the larger structures within the human organism, are also "cuttable" or "dissectable" into smaller pieces. These smaller pieces are technically called the subatomic particles. This is because they exist "under" (*sub-*) the higher level of the entire "atom." The three most common types of subatomic particles are the *protons* (**PROH**-tahns), *neutrons* (**NEW**-trahns), and *electrons* (**e-LEK**-trahns). Within the carbon atom, for instance, there are 6 protons, 6 neutrons, and 6 electrons (Figure 3.1, B).

Each atom contains a *nucleus* (**NEW**-klee-us) – a rounded, centrally located "kernel" (*nucle*) or core. [**Study suggestion:** Think of the nucleus as a hard, round gumball.] The protons are subatomic particles with a positive (+) charge in the nucleus, and they are often accompanied by neutrons which, as their name suggests, are electrically "neutral" (having a charge of 0). Circling at high speed around the nucleus is a *cloud of electrons*, which are tiny, negatively charged (−) particles also having some of the characteristics of waves. [**Study suggestion:** Consider the electron clouds as a sticky swirl of cotton candy surrounding the nucleus gumball.]

SUBATOMIC   =   (Protons + Neutrons)   +   (Orbiting Electrons)
PARTICLES   THE NUCLEUS   FUZZY CLOUD
                                                AROUND NUCLEUS

Since the number of positively charged protons normally equals the number of negatively charged electrons, the net electrical charge of most atoms is zero.

Subatomic
particles 1

Chemical Bonds: Builders of Order and Molecules

We have shown how even the individual atom is dissectable into smaller subatomic particles, which exist at a lower level. But we can also work from the higher end, and link individual atoms together to create *molecules*. A molecule is a "little mass" (*molecul*) of matter that consists of two or more atoms held together by *chemical bonds*.

A chemical bond is a sharing or transfer of outermost electrons between atoms. And as a result of this bonding (sharing or transfer of electrons), the characteristics of the formerly separate atoms are changed into the new characteristics of the molecule. In addition, the process of chemical bonding

Subatomic
particles 1

creates precise new patterns or arrangements of atoms in the body, thereby building Biological Order.

ORGANIC MOLECULES VERSUS INORGANIC MOLECULES AND THEIR BONDING

Molecules created by bonding are of two broad types – *organic* (or-**GAN**-ik) *molecules* versus *inorganic* (**IN**-or-**gan**-ik) *molecules*. The organic molecules literally make up most of the body "organs," which are rich in bonded carbon (C) atoms. The heart, lungs, brain, and other organs, then, all primarily consist of highly orderly arrangements of organic (carbon-containing) molecules.

The organic molecules have a chemical "backbone" that mainly includes either *C–C* (*carbon–carbon*) or *C–H* (*carbon–hydrogen*) *bonds*. These types of bonds are usually classified as *nonpolar* (**nahn-POH**-lar) *covalent* (koh-**VALE**-ent) *bonds*. A covalent bond always involves a sharing "together" (*co-*) of outermost electrons between atoms. An important follow-up question to ask is, "Do the atoms share these outermost electrons *equally*, or do they share them *un*equally?"

If the electrons in the chemical bond are shared equally between atoms, then a nonpolar covalent bond results. Something that is nonpolar does "not" (*non-*) have any *poles* or areas of opposite electrical charge. When electrons, which have a negative (−) charge, are shared equally between two atoms, their negative charges are also shared equally between the atoms. Thus, no *electrical poles* – areas of opposite positive (+) or negative (−) charges – are created.

Consider the chemical bonding in our familiar glucose molecule (Figure 3.2, A). The C–C bonds and C–O (oxygen) bonds forming its six-sided ring structure are all considered nonpolar covalent. Thus, the electron clouds around each pair of shared electrons in the ring can be conveniently modeled as sticky cotton candy pulled halfway between the two atoms being bonded.

The bonding arrangement is markedly different within the water or H_2O molecule (Figure 3.2, B). Water is an inorganic molecule – one that does "not" (*in-*) contain any "carbon" (*organ*) atoms. Recollect that the carbon atom tends to share its outermost electrons equally with other atoms. The O (oxygen) atom, however, is quite another story! Oxygen tends to be an electron "hog," pulling electrons away from other atoms, and into a closer orbit around itself.

This is what happens when it bonds with hydrogen, creating water. The two *H–O* (*hydrogen–oxygen*) bonds are classified as being *polar* (**POH**-lar)

Subatomic
particles 2

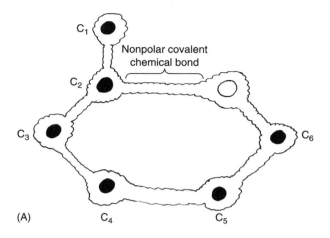

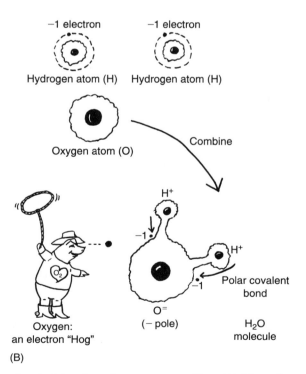

Fig. 3.2 Covalent chemical bonds: polar versus nonpolar. (A) Nonpolar covalent chemical bonds within the glucose molecule. (B) Polar covalent O–H bonds within a water molecule.

covalent. A polar covalent chemical bond is a bond in which the outermost electrons are shared *un*equally between the participating atoms, resulting in a *negative pole* and a *positive pole*. The negative pole is the end where the electrons are pulled more closely, resulting in a net negative (−) charge. Conversely, the positive pole is the end where the electrons have been pulled farther away, leaving it with a net positive (+) charge. [**Study suggestion:** Look back at the anatomy of the atomic nucleus (Figure 3.1). Why does pulling an outer electron away create a net positive charge?]

Covalent bond summary

We can state some simple equations summarizing covalent chemical bonds:

Subatomic
particles 3

$$
\begin{aligned}
\text{COVALENT BONDS} &= \text{THE SHARING OF ELECTRONS} \\
& \text{BETWEEN ATOMS} \\
\textit{NONPOLAR} \text{ COVALENT BONDS} &= \textit{EQUAL} \text{ SHARING OF ELECTRONS} \\
& \text{versus} \\
\textit{POLAR} \text{ COVALENT BONDS} &= \textit{UNEQUAL} \text{ SHARING OF ELECTRONS}
\end{aligned}
$$

Ions: First as Bonds, Then as "Goers-to"

In certain types of bonds, things go far beyond an unequal sharing of electrons. Look at the case of *ionic* (eye-**AH**-nik) *bonds*. Ionic bonds are chemical bonds that result from the complete transfer of one or more outermost electrons between atoms, such that *ions* (**EYE**-ahns) are produced. An *ion* is an atom that has either a net excess or a deficiency of electrons, so that it is electrically charged.

Subatomic
particles 4

$$
\begin{aligned}
\textit{IONIC} \text{ CHEMICAL BONDS} &= \textit{COMPLETE TRANSFER} \text{ OF ELECTRONS} \\
& \text{BETWEEN ATOMS, CREATING } \textit{IONS}
\end{aligned}
$$

A fine example is provided by *sodium chloride* (**KLOR**-eyed), symbolized as *NaCl*. When the sodium (Na) atom gets near a *chlorine* (Cl) atom, it tends to transfer one of its electrons to the chlorine. The net result is an ionic bond between Na and Cl. (Study Figure 3.3.)

The Na end has a net positive (+) charge, while the Cl end has a net negative (−) one. When enough of these ionic NaCl bonds are formed, an elegant *crystal lattice* structure of a salt cube results.

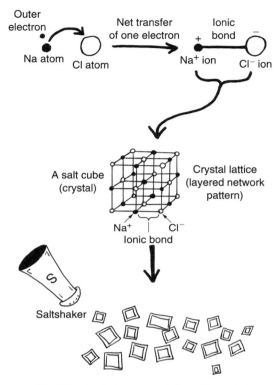

Fig. 3.3 Ionic bonds and the salt crystal.

CATIONS VERSUS ANIONS: TWO KINDS OF IONS

Let us now perform a little experiment. Grab a battery, which has a pair of *electrodes* (ih-**LEK**-trohds). (Examine Figure 3.4.) The negatively charged electrode is called the *cathode* (**KATH**-ohd). It is the source of a stream of negatively charged electrons. Place the cathode into a tank of water, which is a very good "dissolver" or *solvent* (**SAHL**-vent). At the other end of the large water tank, insert an *anode* (**AN**-ohd). This is the positively charged electrode.

Now dump in a bunch of salt (NaCl) crystals. Salt is a type of *electrolyte* (ee-**LEK**-troh-**light**). An electrolyte is a substance that "breaks down" (*lyt*) into ions when placed into a water solvent, such that the resulting solution can conduct an "electrical current" (*electro-*).

Switch-on the electric current at the cathode end. The process called *electrolysis* (ih-**lek**-**TRAHL**-uh-sis) begins to occur. Electrolysis is the "breakdown of" (*-lysis*) some chemical substance under the influence of

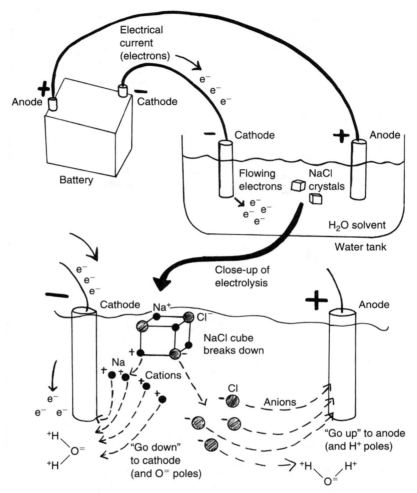

Fig. 3.4   Cations versus anions: ions that "go to" opposite electrodes.

"electricity" (*electro*-). Sodium chloride, being an electrolyte, is readily broken down under the influence of an electrical current flowing through water.
Sodium chloride is broken down into individual Na^+ and Cl^- ions, which are
literally "goers to" one electrode or the other. (This is based on the important
principle that unlike areas of electrical charge attract one another.)

The negatively charged electrons stream out from the cathode and help
attract the positively-charged Na^+ ends of the salt cube. The Na^+ ions are
called *cations* (**CAT**-eye-ahns) because they break away and "go down" from
the dissolving salt cube, and toward the negative *cathode*. The Na^+ ions are
also attracted to the negatively charged $O=$ pole of each water molecule.

The chloride (Cl^-) ions, being negatively charged, are called *anions* (**AN**-eye-ahns) because they "go up" from the chunks of dissolving salt cubes, and toward the positive *anode*. They are also attracted to the positively charged H^+ poles of each H_2O molecule.

In summary,

*CAT*IONS (+)	"Go down to"	THE (−) *CAT*HODE
(*positively* charged ions)		(*negatively*-charged electrode)
	and	
*AN*IONS (−)	"Go up to"	THE (+) *AN*ODE
(*negatively* charged ions)		(*positively*-charged electrode)

Atoms 1

The natural occurrence of millions of dissolved Na^+ cations and Cl^- anions within our body fluids makes the human organism behave as if it were an "internal sea." There are many other types of cations and anions moving around within this "internal sea," as well. The body's internal environment of salty fluids makes it largely electrically charged, because it contains both polar covalent water molecules (with charged poles or ends), and charged cations and anions. In addition, the charged cations, such as Na^+, K^+ or *potassium ions*, and *calcium ions* or Ca^{++}, have important roles to play in making our nerve and muscle cells excitable and responsive to *stimuli* (**STIM**-you-**lie**).

Energy and Metabolism

Ions or "going to" particles are certainly not the only type of chemicals moving around and doing things within the body's internal environment! When ions and molecules are moving around, they must have *energy*. Energy is generally defined as the ability to do work.

THE TWO KINDS OF ENERGY

Energy occurs as two main types – *potential energy* versus *kinetic* (kih-**NET**-ik) *energy*. Potential energy is a form of energy that has the "potential" or possibility of doing work, if it is properly transformed into kinetic energy. Kinetic means "pertaining to" (-*ic*) "movement" (*kinet*). Kinetic energy is the energy associated with particle movement.

Potential energy is energy that is locked up. It is either locked up within some body structure, or locked up when the body is at rest within some position or location. This potential energy is locked up, for example, within the chemical bonds of the glucose molecule. When the chemical bonds are split, however, some of the stored potential energy is released, and converted into kinetic energy. Kinetic energy is alternatively called *free energy*, because it is not stored or locked up. This free or kinetic energy then moves various particles around, and does body work.

ANABOLISM VERSUS CATABOLISM: THE "YIN–YANG" OF METABOLISM

Chemical reactions are interactions between two or more chemicals. For chemicals to interact, of course, they must move around and have kinetic energy. The most important chemical reactions occurring within the living human organism are those of metabolism, a "state of change" (Chapter 1). Traditional Chinese medicine involved the Yin–Yang concept of two opposite but complementary energy forces.

Quite interestingly, metabolism, too, has its own Yin–Yang balance! The two opposing faces of metabolism are *anabolism* (ah-**NAB**-oh-**lizm**) versus *catabolism* (kah-**TAB**-oh-**lizm**). And in each case, their chemical reactions are speeded up by *enzymes* (**EN**-zighms). Enzymes are protein *catalysts* (**KAT**-uh-lists) – substances that speed up chemical reactions without themselves being changed in the process.

1. Synthesis reactions or **anabolism** (ah-**NAB**-oh-lizm) involve a "condition of" (*-ism*) "building up" (*anabol*) something from smaller pieces. (See Figure 3.5, A.) During this process, free energy is usually added or consumed. Let us refer back to our old friend, the individual glucose molecule. Suppose that we eat many glucose molecules, and they enter the liver from our bloodstream. Certain *anabolic* (**an**-ah-**BAHL**-ik) *enzymes* in the liver are activated. These anabolic enzymes then speed up the combining of many individual glucose atoms together (like beads on a string). The result is the synthesis or anabolism of *glycogen* (**GLEYE**-koh-**jen**). The glycogen molecule thus stores a lot of potential energy within its chemical bonds. It is a well-known form of starch-like carbohydrate found within the liver cells and muscle fibers of humans and animals.

2. Breakdown reactions or **catabolism** (kah-**TAB**-oh-lizm) involve a "condition of" (*-ism*) "casting (breaking) down" (*catabol*). (Examine Figure 3.5, B.) That is, a single large molecule, such as glycogen, is acted upon by various *catabolic* (**KAH**-tah-**bahl**-ik) *enzymes*. These enzymes break or cleave the chemical bonds between the component glucose molecules. The results reflect just what the word glycogen literally means – "sweetness" (*glyc*) "producer"

(-*gen*). Free (kinetic) energy is released, and sweet, sugary glucose molecules enter the bloodstream. This might well happen after exercise or fasting.

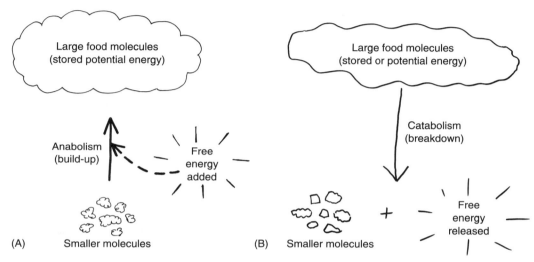

Fig. 3.5   Anabolism versus catabolism: the Dynamic Duo of metabolism. (A) Synthesis reactions or anabolism: "A condition of building-up." (B) Breakdown reactions or catabolism: "A condition of casting (breaking) down."

METABOLISM   =   ANABOLISM   +   CATABOLISM
                 (Building up   (Breaking down
                  molecules)   molecules)

Molecules 1

"Blood Hydrogen Homeostasis": Body Acid–Base Balance

During our discussion of ionic bonding, we brought up the topic of salts, such as common table salt – sodium chloride (NaCl). In general,

$$ACID + BASE = SALT$$

To thoroughly understand the nature of the important salts or electrolytes (like NaCl) in our body fluids, therefore, we need to know a little bit about *acids* and *bases*.

Providing a specific example, we have:

$$\begin{array}{ccccccc} \text{HCl} & + & \text{NaHCO}_3 & = & \text{NaCl} & + & \text{H}_2\text{CO}_3 \\ \text{(Hydrochloric} & & \text{(Sodium} & & \text{(Common} & & \text{(Carbonic} \\ \text{acid)} & & \text{bicarbonate, a weak} & & \text{table salt)} & & \text{acid)} \\ & & \text{base)} & & & & \end{array}$$

The above is often called the *intestinal* (in-**TES**-tih-**nal**) *neutralization* (**new**-tral-ih-**ZAY**-shun) *equation*. The reason is that this particular chemical reaction helps neutralize excess stomach acid that often leaks into the *small intestine*.

ACIDS AS HYDROGEN ION DONORS

Molecules 1

An acid is a *hydrogen ion* or *H^+ ion* donor. An acid, such as *hydrochloric* (**high**-droh-**KLOR**-ik) *acid*, HCl, has a tendency to "donate" (give off) its hydrogen (H^+) ion. So,

$$\text{HCl} = \text{H}^+ + \text{Cl}^-$$

Now, individual H^+ ions, themselves, are very dangerous to the body in high concentrations – Why? The reason is that the H^+ ions are highly reactive and *corrosive* (koh-**ROH**-siv), tending to disintegrate or wear away anything they touch! A *strong acid*, such as HCl, is one that donates *many* H^+ ions, so that it is *very* corrosive! This is just fine within the stomach, where HCl disintegrates and helps chemically digest eaten food. But within the small intestine, a very high H^+ ion concentration may wear away the wall and create *intestinal ulcers*.

BASES OR ALKALI AS HYDROGEN ION ACCEPTORS

Molecules 2

In direct contrast to the acids are the *bases* and *alkali* (**AL**-kah-**lie**). A base is an *acceptor of H^+ ions* (usually donated or given off by an acid). A *weak base*, such as *sodium bicarbonate* (buy-**CAR**-buh-**nayt**) or *NaHCO$_3$*, is a base that accepts just a few H^+ ions. Sodium bicarbonate (NaHCO$_3$) is produced by many cells in the liver and pancreas. It enters the small intestine from these sources, where it reacts with the dangerous strong acid, HCl.

The Na atom from the NaHCO$_3$ combines with the Cl from HCl, such that NaCl and H$_2$CO$_3$ result. The H$_2$CO$_3$ is a chemical formula for *carbonic* (**car**-

BAH-nik) *acid*. Carbonic acid is a *weak acid*, because it donates relatively few H^+ ions. Thus, the intestinal neutralization equation starts out with a strong acid (HCl) and eventually substitutes it with a weak acid (H_2CO_3). In this way, the lining of the intestines is relatively safe, because the strong, corrosive acid (HCl) has effectively been "neutralized" – rendered less dangerous or harmful by being replaced with a much weaker acid.

Alkali as very strong bases

We have said that sodium bicarbonate ($NaHCO_3$) is a weak base. The alkali, in contrast, are very *strong bases*, because they can accept many H^+ ions. Some common alkali include *lye* and *ammonia*. Lye largely consists of the alkali (strong base) called *sodium hydroxide* (high-**DRAHKS**-eyed), *NaOH*. And ammonia, symbolized as NH_3, is a strong base or alkali formed in the body from the catabolism of many nitrogen-containing substances, such as proteins and *amino* (ah-**MEE**-noh) *acids*. Ammonia is very poisonous, and is quickly converted into *urea* (**you-REE**-ah) by the liver.

Besides being potent acceptors of H^+ ions from acids, both lye (NaOH) and ammonia (NH_3) are extremely powerful detergents and dissolvers of greasy, lipid, or "fatty" material. [**Study suggestion:** Ask yourself the following question, "Well, if NaOH and NH_3 are such good acceptors of H^+ ions, then why don't we just drink a large quantity of lye and ammonia, so that they will very effectively neutralize any acid that happens to leak into the small intestine from the stomach? Wouldn't they neutralize excess acid a lot more effectively than sodium bicarbonate ($NaHCO_3$), which is just a weak base?"]

THE CONCEPT OF pH

Besides potentially eating away at the insides of our stomach and intestines, the concentrations of acids and bases within our body fluids critically affect the chemical environment both around and within our cells. Each of the enzymes (protein catalysts) that speed up metabolism, for instance, function within a chemical environment filled with millions of hydrogen (H^+) ions. Each enzyme has what is called an *optimum pH*, representing the H^+ ion concentration where it functions most efficiently, as well as a *permissible range of pH*, which the enzyme must have if it is to function at all!

Expressing H$^+$ ion concentration as pH

Since the concentration of H$^+$ ions is so critical for the functioning of our body's metabolism, we should be able to measure and express it in appropriate units. The concentration of hydrogen ions can be measured, for example, as the number of grams of H$^+$ ion present per liter of fluid (such as the blood). The chief problem with these units, however, is the extremely small numbers of hydrogen ions involved!

We frequently use a method called *scientific notation* for such amazingly small numbers. A *neutral solution or fluid*, such as pure water, is one that is neither acid nor base. Pure water (a neutral fluid) has a hydrogen ion concentration of 1×10^{-7} g of H$^+$/liter. Expressed in decimals, we have 0.0000001, or in fractions, only 1/10 millionth of a gram of H$^+$ ion/liter of pure H$_2$O!

This tiny H$^+$ ion concentration presents a huge mind-boggling problem for us in trying to understand and picture it. As a way of simplifying the situation, we convert the hydrogen ion concentration into the *pH unit*. The pH designation is an abbreviation meaning "*potential of Hydrogen*."

A NEUTRAL FLUID   Has pH $= 7.00$ or 1/10 millionth (10^{-7} of a
(neither acid nor base):   gram of H$^+$ ion) (note *7* decimal places to
the left of 1 in 0.0000001)

This is our starting point. We know that a pH of 7 is 1/10 millionth of a gram of H$^+$, or 0.0000001 (note 7 decimal places to the left of 1). If there is *more than* 1/10 millionth of a gram of H$^+$ present, we have an *acid*. This is because an acid, or H$^+$ ion donor, certainly has a greater hydrogen ion concentration than does pure water!

A H$^+$ ion concentration of 1 *millionth* of a gram, 1×10^{-6} or 0.000001 (note *6* decimal places to the left of 1), has a pH $= 6$. This makes it an *acid*! [**Study suggestion:** Would you rather inherit 1 millionth of a fortune, or 1/10 millionth of it? Isn't the 1 millionth more money (or H$^+$ ions)?]

AN ACID: Has pH $= 6$.*something or less*, <u>not</u> 7!

A H$^+$ ion concentration of 1/*100 millionth* of a gram, 1×10^{-8} or 0.00000001 (note *8* decimal places to the left of 1), has a pH $= 8$. This makes it a *base*! Certainly, a base would have a lower hydrogen ion concentration than even pure water! It pretty much *has* to be low, because a base is an *acceptor* of H$^+$ ions, isn't it?

A BASE OR ALKALI: Has pH =7. *something, 8. something, or higher, <u>not</u>* *just 7!*

Yes, it seems pretty weird, but:

ACIDS: Have *low* pH, but a *HIGH* H^+ ion concentration (because they have plenty and donate H^+ ions)

BASES: Have *HIGH* pH, but a *low* H^+ ion concentration (because they are "needy" and accept donated H^+ ions)

SAMPLE pH SCALE OF FAMILIAR FLUIDS

To help you visualize changes in pH and hydrogen ion concentration, consult the sample pH scale of familiar fluids displayed in Figure 3.6. It may assist you in noting that the word acid literally means "sour"! So drinking **sour** things with a lot of **acid** and **hydrogen ions** (such as lemon juice), make you **pHeel (feel) low** and sick and pucker your lips! But drinking some beers with a lot of **base** make you **pHeel (feel) high** and **hops!** (Bases in general taste bitter and feel soapy.)

ACID–BASE BALANCE

In general, we say that a person is in a state of *acid-base balance* or *"blood hydrogen homeostasis"* if their blood pH is changing slightly around its setpoint value of about 7.4. [**Study suggestion:** Does this make blood slightly acid, or slightly basic?]

The blood pH can increase to an upper normal limit of about 7.5, and decrease to a lower normal limit of about 7.3. We can then legitimately say that a state of acid–base balance or blood hydrogen homeostasis is being maintained over time. (Check Figure 3.7.)

Subatomic particles 5

Acidosis Versus Alkalosis: Upsets in the Balance

Whenever blood hydrogen ion concentration rises beyond its upper normal limit of about 7.5, then a state of *alkalosis* (**al**-kah-**LOH**-sis) is said to exist. Alkalosis is an "abnormal condition of" (*-osis*) too much base or alkali, or

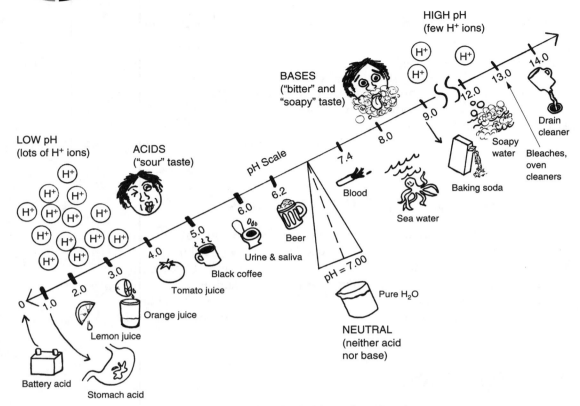

Fig. 3.6 Some familiar fluids on the pH scale.

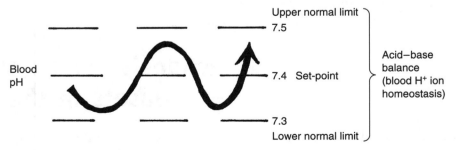

Fig. 3.7 Homeostasis of blood H^+ ion concentration: A condition of body acid–base balance.

not enough acid. Consider, for example, what would happen to Baby Heinie if he got into his mother's cupboard and drank a bottle of ammonia! Yes, the poor kid would have a dangerous case of *ammonia poisoning*, but also alkalosis! In this case, it is called *metabolic alkalosis* or an *alkali overdose* (Figure 3.8, A).

Subatomic particles 1

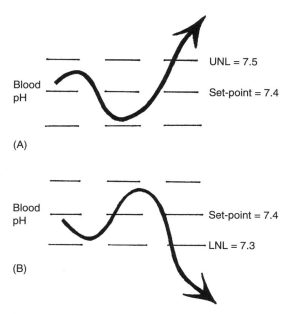

(A)

(B)

Fig. 3.8   Acidosis versus alkalosis. (A) Metabolic alkalosis or alkali overdose. (B) Metabolic acidosis or acid overdose.

Whenever blood hydrogen ion concentration falls below its lower normal limit of about 7.3, then a state of *acidosis* (**ah**-sih-**DOH**-sis) is said to exist. Acidosis is an "abnormal condition of" (-*osis*) too much acid, or not enough base or alkali. Contemplate what would happen if Baby Heinie got into his mother's cupboard and drank a huge bottle of sour vinegar! The result would be *metabolic acidosis* or *acid overdose* (Figure 3.8, B).

Subatomic particles 2

Quiz

Refer to the text in this chapter if necessary. A good score is at least 8 correct answers out of these 10 questions. The answers are listed in the back of this book.

1. First proposed the existence of atoms:
   (a) Socrates
   (b) Sophie Tucker
   (c) Hippocrates
   (d) Democritus

2. The atom is not really "uncuttable" because:
   (a) Ions only occur in certain substances
   (b) Splitting atoms reveals their subatomic particles
   (c) No one has developed the necessary technology to try cutting them
   (d) Atoms do not really exist!

3. Nonpolar covalent bonds:
   (a) Are mainly found in association with acids
   (b) Involve the net transfer of electrons between atoms
   (c) Create molecules with poles having opposite charge
   (d) Represent the equal sharing of outermost electrons between atoms

4. NaCl is classified as an electrolyte, since:
   (a) Its crystals readily break down into ions that can help water conduct electricity
   (b) Its salty nature makes its taste "electrical"
   (c) NaCl is decomposed into sodium anions and chloride cations
   (d) The crystal lattice has a C atom at each of its corners

5. Kinetic energy represents:
   (a) Stored energy that could potentially do work
   (b) Free energy associated with particle movement
   (c) Calories locked up into chemical bonds
   (d) Cold temperature conditions that can never be changed

6. The reaction, Glucose + Fructose = Sucrose, is an example of:
   (a) Catabolism
   (b) Electrolysis
   (c) Anode discharge
   (d) Anabolism

7. Acid + Base = _____ :
   (a) Alkali
   (b) Complex carbohydrate
   (c) Cl^-
   (d) Salt

8. Important H^+ ion acceptors:
   (a) Catabolic enzymes
   (b) Alkali
   (c) Acids
   (d) Most fruit juices

9. A fluid with a hydrogen ion concentration of 0.0000000001 g/liter has a pH of:
   (a) 6
   (b) 8
   (c) 10
   (d) 12

10. Severe vomiting greatly reduces the amount of HCl left in the stomach and creates:
   (a) Intestinal ulcers
   (b) Metabolic acidosis
   (c) Respiratory alkalosis
   (d) Metabolic alkalosis

Body-Level Grids for Chapter 3

Several key body facts were tagged with numbered icons in the page margins of this chapter. Write a brief summary of each of these key facts into a numbered cell or box within the appropriate *Body-Level Grid* that appears below.

Anatomy and *Biological Order* Fact Grids for Chapter 3:

A

*SUBATOMIC
PARTICLE
Level*

1

ATOM
Level

1

Physiology and Biological Order **Fact Grids for Chapter 3:**

MOLECULE
Level

1

Function and *Biological Order* **Fact Grids for Chapter 3:**

SUBATOMIC
PARTICLE
Level

1	2

3	4

5

ATOM
Level

1

MOLECULE
Level

1	2

Function and *Biological Disorder* Fact Grids for Chapter 3:

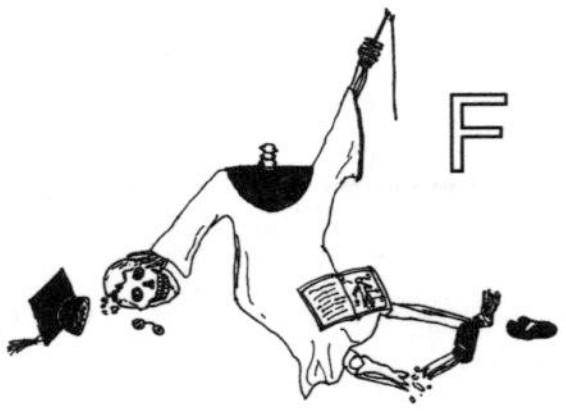

SUBATOMIC
PARTICLE
Level

1	2

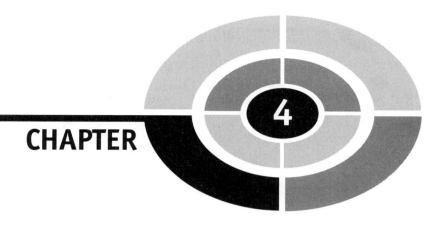

The Cell: A Tiny Arena of Life

We have now climbed up out of the Body Basement. You know – that was the Chemical Level (Chapter 3). Remember that we have just Plain Body Functions at the level of the subatomic particles, atoms, molecules. This is a fact, of course, because it is only until we reach the Cell Level that Plain Body Functions finally become physiology.

The Cell Level: Microanatomy Becomes Microphysiology

In our companion volume, *ANATOMY DEMYSTIFIED*, we read about *microanatomy* (**MY**-kroh-uh-**NAT**-uh-me), also called *microscopic* (**my**-kroh-**SKAHP**-ik) *anatomy*.

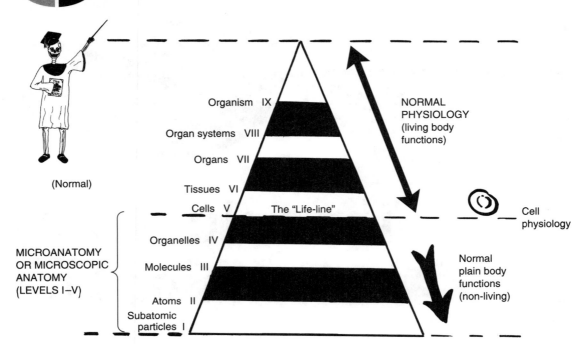

Technically speaking, microanatomy would include the subatomic particles, atoms, molecules, organelles, and cells making up the structure of the human body. (Although the non-detailed characteristics of many tissues are visible to the naked eye, the individual cells making up these tissues are usually studied with the microscope.)

The cell is the *highest* level of *microanatomy*, but at the same time it is the *lowest* level of *microphysiology* (**my**-kroh-fih-zee-**AHL**-uh-**jee**). We mean that the cell's structures are so small that they must be viewed under a microscope, but the functions of the entire cell are now officially called physiology, because the cell is alive. The microphysiology of the cell is more often described by another phrase – *cell physiology*.

CELL   =   MICROPHYSIOLOGY   =   FUNCTIONS
PHYSIOLOGY   AT THE CELL   CARRIED OUT BY
LEVEL   THE LIVING CELL

An interesting related question is, "What about the cell *organelles*?"

THE CELL ORGANELLES "DO THEIR THING" FOR THE CELL'S SURVIVAL

An organelle is literally a "tiny" (*-elle*) "organ"-like structure that carries out a certain specialized function within a cell. An overview of some of the major cell organelles is provided in Figure 4.1. Even though the entire cell is alive, the numerous organelles within the cell are *not* themselves alive! Rather, these examples of cell anatomy perform critical functions or each "do their own thing" to promote the survival and physiology of the entire cell as our fundamental unit of life.

Organelle 1

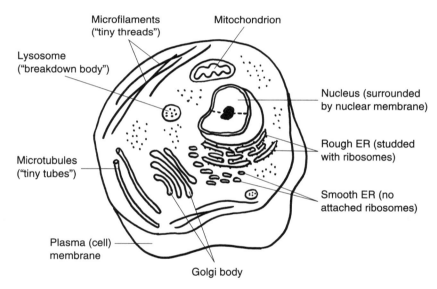

Fig. 4.1. The major organelles within the living human cell.

The Cell Membrane: A Selectively Permeable Boundary

When you think of cell organelles, perhaps the first one that comes to mind is the *cell or plasma* (**PLAZ**-muh) *membrane*. This is because the cell (plasma) membrane is the *selectively permeable* (**PER**-me-ah-**bl**) *barrier* that limits and

Organelle 2

Molecules 1

surrounds the cell. A selectively permeable membrane is a thin barrier that allows only certain types of particles to "cross through" or *permeate* (**PER**-me-ate) it. At the same time, the membrane limits or prevents the passage of other particles.

Figure 4.2 provides a detailed view of what is called the *fluid mosaic* (moh-**ZAY**-ik) *model* of the cell membrane. A mosaic, in general, is a complex pattern of objects having different sizes, shapes, and colors. [**Study suggestion:** Closely examine a thin mosaic surface, such as on a floor or wall. Do you see a definite repeating pattern of different sizes, shapes, and colors within this mosaic? The cell or plasma membrane is now thought to have a structural pattern that is somewhat similar.]

The most abundant chemical component of the cell membrane is its *phospholipid* (**fahs**-foh-**LIP**-id) *bilayer* (**BY**-lay-er). This is a "double" (*bi-*) layer of *lipid* (**LIP**-id) or "fatty" molecules that have long *fatty acid tails* and a charged, polar, *phosphate-*(**FAHS**-fayt) *nitrogen* head group. Observe in Figure 4.2 that one layer of the electrically charged (polar) phosphate-nitrogen heads are oriented *outward* from the cell membrane, in contact with the watery *extracellular* fluid, and the other layer of charged head groups is

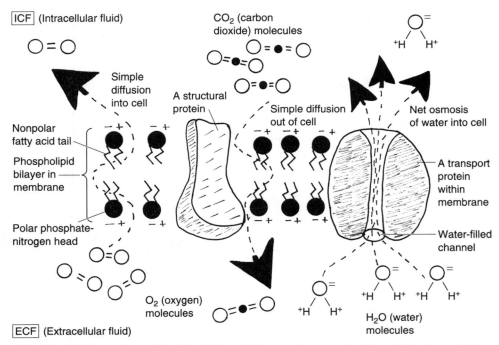

Fig. 4.2. The fluid mosaic model of the cell membrane and movement of particles through it.

oriented *inward* from the membrane, in contact with the watery *intracellular* fluid.

SIMPLE DIFFUSION AND OSMOSIS: THE PRINCIPLE OF "LIKE DISSOLVES LIKE"

In determining *which* types of particles cross through (permeate) the cell membrane, and *how* they pass through it, the *principle of "like dissolves like"* is very essential. To be precise,

THE PRINCIPLE OF "LIKE DISSOLVES LIKE":	Chemicals with similar or "like" characteristics of bonding tend to mix or dissolve well together. So, charged particles tend to mix well with one another; and uncharged, nonpolar particles tend to mix well with each other. Just don't try to mix the *charged* particles with the *uncharged* ones!

Let us consider, for instance, the uncharged, nonpolar covalent, fatty acid tails of the phospholipid bilayer in the cell membrane. Being electrically uncharged, this bilayer of fatty acid tails does not allow charged particles, such as polar covalent H_2O molecules, or ions such as Na^+, to mix with it. The fatty acid tails do, however, let uncharged *oxygen* (O_2) molecules mix, dissolve, and *diffuse* (dih-**FUSE**) or randomly "scatter" through them. Likewise, they allow uncharged *carbon dioxide* (die-**AHKS**-eyed) or CO_2 molecules to mix and diffuse across the cell membrane.

Molecules 1

Simple diffusion is the random (chance) scattering of particles from a region where their concentration is high, to a region where their concentration is low. Oxygen (O_2) molecules usually have a relatively high concentration outside most of our cells, and a relatively low concentration within our cells. [**Study suggestion:** Ask yourself, "Why does oxygen have such a low concentration within most of our body cells? What happens to the O_2 molecules, when they diffuse in? Are they left alone, or are they used by cell metabolism for something special?"]

Therefore, there is usually a *net* (overall) simple diffusion of O_2 molecules across the fatty acid bilayer, from the outside of the cell to the inside of the cell. Conversely, the net diffusion of CO_2 molecules is in the opposite direction (from the inside of the cell to the outside of the cell), because CO_2 is at higher concentration within the cell. [**Study suggestion:** Think about this question: "Where is the CO_2 coming from, inside of the cell? What is going on to produce it?" We will come back to this problem later in the chapter.]

Glancing back at Figure 4.2, observe the presence of both *structural proteins* as well as *transport proteins*, occurring here and there within the cell membrane. Note that the transport proteins have a water-filled channel that extends all the way through them. These channels through transport proteins are the way that the charged, polar H_2O molecule can diffuse through the membrane. (They cannot mix with and diffuse through the nonpolar, uncharged fatty acid tails of the phospholipid bilayer, of course, because this would violate the "like dissolves like" principle.)

Molecules 2

Osmosis (ahs-**MOH**-sis) is literally a "condition of" (-*osis*) "thrusting" (*osm*). Osmosis is defined as the simple diffusion of water (H_2O) molecules, from a region of high water concentration to a region of low water concentration. In the cell membrane, such simple diffusion of water occurs through the channels in the transport proteins.

Water can, of course, pass through the channels in both directions. But if the water concentration is higher in the ECF (extracellular fluid), then there will be a *net* (overall) osmosis of water *into* the cell. If the water concentration is higher in the ICF (intracellular fluid), then there will be a net osmosis of water *out of* the cell. [**Study suggestion:** What do you predict will happen to the size of the cell if there is a net osmosis of water into it? What if there is a net osmosis out of it?]

The idea of a "concentration gradient"

You may have noted something in common about our discussions of both simple diffusion (random scattering of particles other than water) and osmosis (the diffusion of water molecules only). This is the fact that *net* diffusion or osmosis always occurs "downhill." We mean that there is always a net or overall movement of particles down their so-called *concentration gradient* (**GRAY**-dee-**unt**). The concentration gradient is literally a "grade" or "hill" of concentration that exists between two different areas (as between the ECF and ICF).

NET DIFFUSION OR   Always occurs *"downhill"* (from an area where
OSMOSIS: the concentration of the moving particles is
                              higher, to an area where their concentration is
                              lower.)

[**Study suggestion:** Visualize a steep hill or gradient. A large bunch of randomly moving balls (chemical particles) are at the top of the hill, while a small number are at the bottom. Purely due to chance, then, in what direction will there tend to be a net or overall movement of these balls – up the hill, or down the hill? Confirm your answer by looking back at Figure 4.2. In what direction is the net diffusion of O_2 molecules and CO_2 molecules, and the net osmosis of H_2O molecules, through the plasma membrane?]

MEDIATED TRANSPORT PROCESSES: THE ROLE OF PROTEIN ''CARRIER'' MOLECULES

Many particles are just too large to diffuse through either the phospholipid bilayer, or the water-filled protein channels, within the cell membrane. Such particles have their transport helped or *mediated* (**ME**-dee-**ay**-ted) by something in the "middle" (*medi*) of the membrane. These *mediators* (**ME**-dee-**ay**-ters) in the "middle" are called *protein "carrier" molecules*. A protein carrier is a molecule that has one or more *binding sites* where certain chemical particles can attach or "bind" to it. (This is somewhat like docking sites for boats of particular sizes.)

Molecules 3

When the carried chemical particle attaches to the binding site, the protein carrier changes its shape or *conformation* (kahn-for-**MAY**-shun). As a result, the protein carrier shifts around within the cell membrane, such that the binding site (with its attached chemical particle) now becomes positioned on the *opposite* side of the membrane. The carried particle is then released from its binding site, having effectively crossed through the cell membrane.

Facilitated diffusion of glucose

We have repeatedly mentioned the fairly large glucose molecule, which has a hexagon-shaped ring structure. Glucose enters most of our cells from the blood or ECF by a process called *facilitated* (fuh-**SIL**-uh-**tay**-ted) *diffusion*. When some process is facilitated, its progress is made "easier" (*facilit*).

Molecules 4

Facilitated diffusion is the diffusion of some particle (such as glucose) that is "facilitated" or made easier by the help of a protein carrier molecule within the cell membrane. Perhaps our understanding of this facilitated diffusion process can be helped by making an appropriate metaphor. Suppose that the protein carrier molecule for glucose is a nice soft, fluffy, white pillow (Figure 4.3, A). And pretend that the pillow is on Baby Heinie's bed (the cell membrane). Now, Baby Heinie's Mom tells him, "Eat your peas, or else!" The kid refuses to eat his peas (green vegetables which Mom knows are good for him), so he goes to bed with his dinner plate full of a high concentration

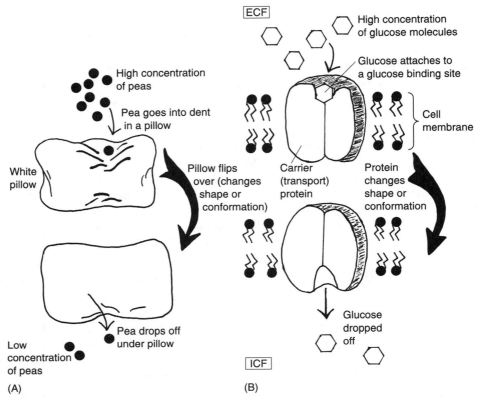

Fig. 4.3. Facilitated diffusion of glucose: a story about a "little pea" and its "pillow." (A) "Peas in the pillow": A visual metaphor. (B) The facilitated diffusion of glucose into a cell.

of peas. When he falls asleep, the plateful of peas falls onto his pillow. At least one of them lodges into a tiny dent in the pillow surface. This dent can be thought of as a "binding site" for the little pea. Now, there is a high concentration of spilled peas on the top of Baby Heinie's pillow, but a very low concentration of peas under his pillow. Baby Heinie has a restless night, tossing and turning in his bed, and socking and punching at his pillow, thereby changing its shape or conformation. Eventually, the surface dent with its pea becomes flipped around onto the bottom of the pillow, where it is released and falls under the pillow. Thus, there has been a "mediated transport" of the little pea by a "facilitated diffusion" using Baby Heinie's pillow!

Let us compare this fanciful scenario to the actual mediated transport of glucose (Figure 4.3, B). Glucose molecules are at a relatively high concentration within the interstitial fluid outside the tissue cell, because Baby Heinie has just eaten his peas – which are sweet and rich in glucose! An individual glucose molecule randomly hits and fits into a *glucose binding site* on a carrier protein. The protein carrier changes its shape or conformation, such that the glucose binding site becomes oriented on the inner face of the cell membrane. The glucose molecule thus moves across the membrane by a facilitated diffusion process, with the carried glucose molecule finally being released into the ICF (intracellular fluid). Note that this is still a net diffusion process, because the glucose uses the help of a protein carrier in the membrane to move down its concentration gradient, from a relatively high glucose concentration in the ECF, to a relatively low concentration in the ICF. [**Study suggestion:** Ask yourself, why does glucose exist at a relatively low concentration within most of our body cells? Hint: What usually happens to it, once it enters the intracellular fluid and cytoplasm by facilitated diffusion?]

Active transport of sodium and potassium ions

We have not really considered the movement of small ions, such as sodium ions (Na^+), and *potassium* (puh-**TAH**-see-**um**) *ions*, symbolized as K^+, across the plasma membrane. Like the movement of charged, polar H_2O molecules during osmosis, these ions pass through channels in special membrane proteins (Figure 4.4, A). Sodium ions are present at a higher concentration in the ECF, so they have a net simple diffusion *into* the cell, through special *sodium channels* in particular membrane transport proteins. Conversely, potassium ions are present at a higher concentration within the ICF. Thus, they have a net simple diffusion *out of* the cell, through special *potassium channels* in certain membrane transport proteins.

The continuous operation of these two ion diffusion systems presents the cell with a special problem: after sodium and potassium ions have been diffusing "downhill" for a long enough time, how is a *diffusion equilibrium* of Na^+ and K^+ avoided? In other words, how does the cell prevent Na^+ from eventually having a concentration within the ICF that is exactly equal to that in the ECF? The same question can be asked about the relative concentrations of K^+ ions inside of, and outside of, the cell.

The answer is a short one – the existence of a *sodium–potassium exchange pump*. The sodium–potassium exchange pump is the most common type of *active transport system* that operates across the cell membrane. Active transport requires the use of free energy obtained from enzymes *actively* splitting a molecule called *ATP*. Active transport, then, is defined as the active movement or "pumping" of particles "uphill," from a region where their

Molecules 5

concentration is low, up to a region where their concentration is high. Visualize the Na^+ ions, for example, as moving *up* (not down) their concentration gradient. (This is like trying to actively roll a boulder up a steep hill.)

The specific case of active transport using a sodium–potassium exchange pump is illustrated in Figure 4.4 (B). For simplicity, we will once again model the protein carrier involved as a big, soft, white, fluffy pillow. "Baby Heinie, eat your peas and carrots!," Mom orders in complete frustration. Baby Heinie stubbornly refuses to eat his vegetables at dinner, so he takes his plateful of peas and carrots to bed with him. Now, there is already a low concentration of dark peas (Na^+ ions) under Baby Heinie's pillow (within the intracellular fluid), which slipped in from the last time he refused to eat his veggies!

Some peas (Na^+ ions) slide off of Baby Heinie's dinner plate, so there is a higher concentration of peas or Na^+ ions within the interstitial fluid (above the pillow). Nevertheless, 3 peas (Na^+ ions) at a time attach to three tiny dents (binding sites) under the pillow. Since Na^+ or peas is at a high con-

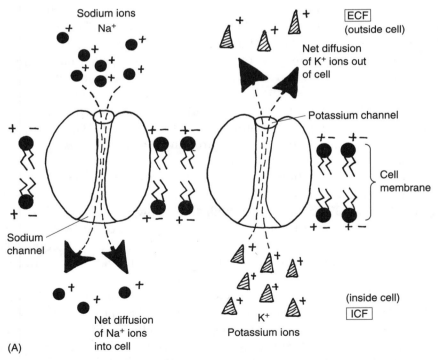

Fig. 4.4 Baby Heinie's "peas" and "carrots": the sodium–potassium exchange pump. (A) Net diffusion of Na^+ and K^+ ions into and out of a cell.

centration above the pillow, Babie Heinie will have to use a lot of free energy from ATP to "pump" (via active transport) the 3 Na$^+$ ions uphill, by changing the shape or conformation of his carrier protein "pillow." As a result of this energy use, 3 Na$^+$ ions are actively pumped back out into the interstitial fluid. Since Na$^+$ ions are *actively transported out of the cell faster* than they *diffuse into the cell* through sodium channels in transport proteins, the Na$^+$ concentration remains lower within the ICF, compared to that in the extracellular fluid.

What about the potassium ions – Baby Heinie's uneaten batch of little orange "carrots"? Well, the kid really hates those carrots! So he already has a high concentration of these uneaten "carrots" (K$^+$ ions) from past dinners, stuffed under his pillow (within the intracellular fluid). Nevertheless, after Baby Heinie's distorted, flipped-around pillow dumps off its 3 peas (Na$^+$ ions) on the outside of the cell, it picks up 2 carrots (K$^+$ ions) from the interstitial fluid. The carrots fit into tiny dents (*potassium ion binding sites*) on the outer surface of the pillow. As the pillow flips back into its original position, these 2 K$^+$ ions are transported down into the intracellular fluid.

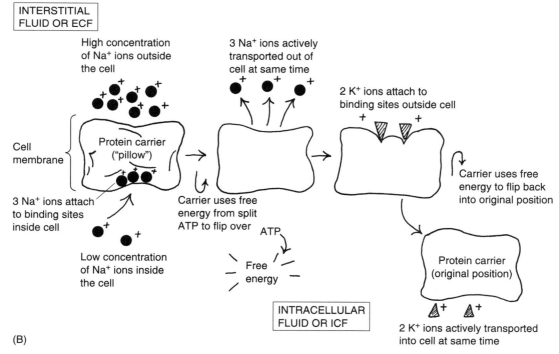

(B)

Fig. 4.4 (continued) (B) A sodium–potassium exchange pump: operation of an active transport system across the cell membrane.

As a result of this total active transport process, 3 Na$^+$ ions are pumped out of the cell for every 2 K$^+$ ions that are pumped into the cell. Therefore, this is a sodium-potassium 3-to-2 *exchange* pump.

SUMMARY OF PARTICLE MOVEMENT ACROSS THE CELL MEMBRANE

In summary,

THE THREE GENERAL TYPES OF PARTICLE MOVEMENT ACROSS CELL MEMBRANE	=	SIMPLE DIFFUSION	+ OSMOSIS +	MEDIATED TRANSPORT

and

MEDIATED TRANSPORT (Use of protein carriers)	=	FACILITATED DIFFUSION (No energy needed)	+	ACTIVE TRANSPORT (Requires use of energy)

Free Energy, Cell Work, and the ATP–ADP Cycle

In addition to active transport across the cell membrane, there are many energy-requiring processes that occur within the cell. We have already mentioned that a critically important molecule, ATP, was split by enzymes and provided the free energy needed by the cell. ATP is an abbreviation for *adenosine* (ah-**DEN**-oh-seen) *triphosphate* (try-**FAHS**-fayt).

THE ATP–ADP CYCLE

The ATP (adenosine triphosphate) molecule is often symbolized as: **A-P~P~P**. The letter, A, of course, is an abbreviation for adenosine. Each letter P denotes phosphate – a phosphorus-containing chemical group. Within ATP, there are "three" (*tri-*) phosphate groups attached to the adenosine. The last two chemical bonds in the ATP are special high-energy bonds. This is indicated by the squiggly line, ~, before each of the last two phosphate groups (~**P**~**P**).

Many cells contain a special type of enzyme called *ATPase* (ATP-ace), literally meaning "ATP splitter" (*-ase*). ATPase enzyme, therefore, acts to split the second high-energy phosphate bond within the ATP molecule. When this bond is split, large amounts of previously stored potential energy is converted into free (kinetic) energy and is used to do work, such as active transport. After losing the end phosphate group, ATP or **A-P~P~P**, becomes *ADP* or *adenosine diphosphate* (**die-FAHS**-fate): **A-P~P**. ADP, therefore, is a reduced version of ATP that contains "two" (*di-*) phosphate groups rather than three.

Molecules 6

When a person eats, say, a candybar or other foodstuff containing a high number of carbohydrate molecules (such as glucose), the individual carbohydrate molecules are eventually broken down by catabolism. The potential energy stored in their chemical bonds is released. The free energy released from catabolism of food molecules is often used to re-attach the end phosphate group back onto ADP, thereby re-creating more ATP. During exercise or fasting, however, the available cell ATP molecules are broken back down into ADP, such that more free energy is released to do the body's work.

This back-and-forth process between ATP and ADP can technically be called the *ATP–ADP cycle*. This energy storage–energy release cycle is a continual process that goes around and around, for as long as the cell lives.

In summary, the ATP–ADP cycle:

ADP or A-P~P + Excess free energy + Phosphate (P) = ATP or A-P~P~P
(Energy storage)

and

ATP split by ATPase enzyme = ADP + Free energy to do cell work

Glucose Breakdown: Is It Going on Within the Cytoplasm, or in the "Thread-Granules"?

With all this talk of energy from ATP, and its use for numerous activities within the cell, it is only natural for us to ask, "If free energy from ATP is so important for the cell, then quite a lot of it must be produced *locally*, right there within the cell, *right*?" Yes, this is quite correct!

GLYCOLYSIS OCCURS IN THE CYTOPLASM

We have already noted that glucose is the major fuel for our body cells. Specifically, it is the *potential energy* stored within the chemical bonds of the glucose molecule, which ultimately serves as our main source of energy.

Since the energy is stored, the bonds within glucose must be broken by the action of certain catabolic enzymes (Chapter 3). The technical term for this action is *glycolysis* (gleye-**KAHL**-ih-sis) – the "process of breaking down" (*lysis*) "sweets or glucose" (*glyc*). The essential chemical steps in glycolysis are outlined in Figure 4.5. Note that one 6-carbon glucose molecule is broken down into two 3-carbon *pyruvic* (pie-**ROO**-vik) *acid* molecules. As a result, enough free energy is released from the broken chemical bonds in glucose to yield a net of 2 ATPs for cell work.

Glycolysis takes place within the *cytoplasm* (**SIGH**-toh-plazm) – the fluid "matter" (*plasm*) of the "cell" (*cyt*). But glycolysis is really inefficient, in that

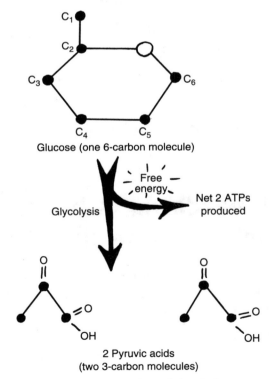

Fig. 4.5 An overview of glycolysis.

the breakdown of 1 glucose molecule results in a net of only 2 ATPs produced. Now, doesn't the cell deserve to get much more ATP energy "bang" out of each of its glucose "bucks"?

CELLULAR RESPIRATION OCCURS IN THE MITOCHONDRIA

Under normal conditions, most of the cells in the body operate within an *aerobic* (air-**OH**-bik) or "living" (*bi*) in "air" (*aer*) environment. As Figure 4.6 reveals, the majority of our cells have a steady supply of oxygen (O_2) being delivered to them from the bloodstream and the interstitial fluid. This oxygen, of course, ultimately comes from the air that we breathe. Hence, our cells mostly operate under aerobic or *oxidative* (**AHKS**-ih-**day**-tiv) – "pertaining to" (*-ive*) "oxygen" (*oxidat*) – conditions.

When both oxygen and glucose are present within the cytoplasm, then the 2 pyruvic acid molecules produced by glycolysis enter into *cellular respiration* (**res**-pir-**AY**-shun). Cellular respiration, then, is just the general phrase for all of the aerobic (oxygen-using) steps in glucose catabolism (breakdown) that occur after glycolysis.

Enter the mitochondria

A

Organelles 1

Whereas glycolysis occurs in the cytoplasm, cellular respiration occurs in the *mitochondria* (**my**-toh-**KAHN**-dree-ah). The mitochondria are literally "thread" (*mito*) "granules" (*chondr*), because sometimes they are long and thin (like threads), while at other times they are round like little grains (granules). Each *mitochondrion* (**my**-toh-**KAHN**-dree-un) is surrounded by an outer membrane plus an inner membrane. The inner membrane is thrown into slender "crests" or "ridges" called *cristae* (**KRIS**-tee). And within the middle of each mitochondrion is a fluid-filled "womb" or *mitochondrial matrix* (**MAY**-tricks).

The mitochondrial matrix is where the 2 pyruvic acids diffuse, after they are produced by glycolysis. Each pyruvic acid takes its turn entering the *Krebs cycle*. Named after the German biochemist, *Hans Krebs*, the Krebs cycle is a repeating cycle of aerobic reactions that break down the pyruvic acids produced by glycolysis. The Krebs cycle rotates twice (once for each of the pyruvic acids fed into it). Along the way, a total of 2 ATPs, several CO_2 molecules, and a number of *hydrogen-carrier molecules* result.

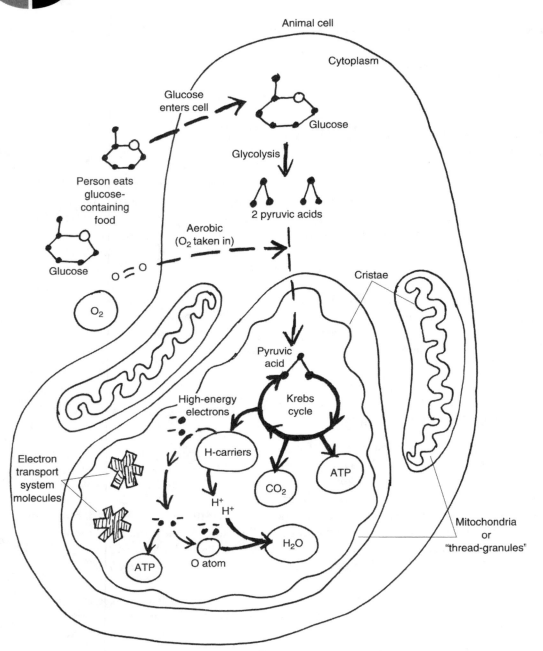

Fig. 4.6 An overview of cellular (aerobic) respiration.

The hydrogen-carrier molecules, once produced by the Krebs cycle, move onto the nearby *electron transport system*. The electron transport system is located along the cristae (inner crests or ridges) of the mitochondrion. The large molecules of the electron transport system, as their name indicates, carry high-energy electrons from H atoms, down to progressively lower and lower energy levels. As the electrons flow or are transported "downhill" (energy-wise), they release considerable amounts of free or kinetic energy along the way.

A net total of 34 more ATPs are eventually produced from this electron transport process. Finally, the transported electrons, now depleted of most of their former energy, are transferred to an oxygen atom. The oxygen atom then combines with two H^+ ions, thereby creating water (H_2O).

The overall equation for cellular (aerobic) respiration, using glucose as the fuel molecule, is:

$$C_6H_{12}O_6 \quad + \quad 6\,O_2 \quad \rightarrow \quad 6\,CO_2 \quad + \quad 6\,H_2O \quad + \quad 36\ ATP$$

Glucose	Oxygen	Carbon dioxide	Water	Free energy

SUMMARY OF AEROBIC GLUCOSE BREAKDOWN

Capsulizing this section, we have:

Glycolysis (occurs in cytoplasm)	2 ATPs/glucose
Cellular respiration (Krebs cycle and electron transport system in mitochondria)	36 ATPs/glucose

Total = 38 ATPs/glucose

Molecules 7

Protein Synthesis: The "Colored Worms" Open and Squirm!

So far, we have talked about the cell (plasma) membrane, the cytoplasm, and the mitochondria. But there are several other key organelles deeply involved in the process of *protein synthesis*, the making of new proteins for the cell. Proteins serve such critical functions as being enzymes (protein catalysts that speed up chemical reactions), structural proteins, transport proteins with channels, and protein carrier molecules.

The most important single place in the cell for protein synthesis is the *cell nucleus*. [**Study suggestion:** You have already heard about another kind of "nucleus" in this book! – Do you remember it? At what level of Body Organization does it exist?]

Somewhat resembling a rounded "kernel" or "nut" (*nucle*), the cell nucleus has its own "skin" (like a peanut)! The "skin" in this case, however, is the *nuclear* (**NEW**-klee-ar) *membrane* surrounding the nucleus. Enclosed within the nuclear membrane are a number of important structures, such as the *chromosomes* (**KROH**-moh-sohms) – "colored" (*chrom*) wormlike "bodies" (-*somes*). (Consult Figure 4.7.) Each of these chromosomes, in turn, contains a tightly coiled *DNA* molecule. There are numerous *genes* (jeans) or sections strung along the DNA molecule.

DNA is an abbreviation for *deoxyribonucleic* (dee-**ahk**-see-**RYE**-boh-new-**KLEE**-ik) *acid*. This name reflects the fact that D*NA* is a *nucleic* (new-**KLEE**-ik) *acid* (*NA*), since it is an "acid" found within the cell "nucleus." The D in the abbreviation stands for *deoxyribose* (dee-**ahk**-see-**RYE**-bohs), a type of sugar "lacking" (*de-*) an "oxygen" (*oxy*) atom.

But in looking at a deoxyribonucleic acid (DNA) molecule, one does not really notice any sugars! Rather, one immediately notices the dramatic *double helix* (**HEE**-liks) or double "spiral" shape. The deoxyribose sugar molecules occur at intervals along each twisted chain of this double helix or twisted spiral.

Specifically, a deoxyribose sugar molecule forms part of the repeating structure of each gene within the DNA double helix. A single gene provides a chemical code for the synthesis of a particular protein. As Figure 4.7 reveals, during protein synthesis, part of the DNA double helix unwinds and puffs out from the surface of the chromosome, almost as if it were a cut-open, squirming earthworm on a fish hook!

DNA CODONS AS CHEMICAL CODE WORDS

A group of *DNA codons* (**KOH**-dahns) becomes exposed along the edges of this chromosome puff. A codon is a chemical "code word" consisting of three nitrogen-containing bases. Each codon is a code word for bringing in a particular type of *amino* (ah-**MEE**-noh) *acid*. A protein is essentially a string of connected amino acids. Therefore, a gene consists of a number of codons, all of them together providing a total code word to the cell for synthesizing a certain protein made of amino acids.

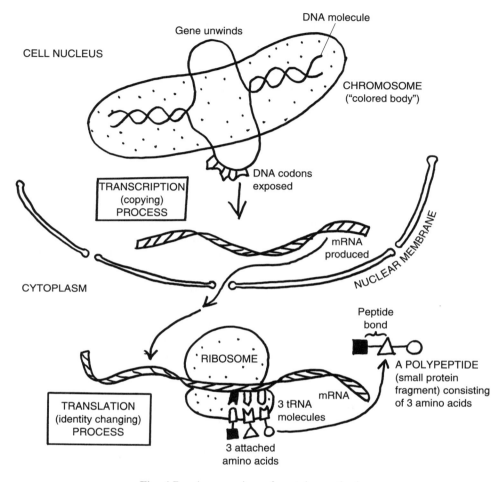

Fig. 4.7.   An overview of protein synthesis.

To summarize this rather involved section:

DNA codon = Chemical "code word" for a certain amino acid;
                                    and

Gene   = A group of codons (chemical "code words") in the DNA
                      molecule

                 = A collection of coded instructions for putting a group of
                      amino acids together in a certain sequence, thereby
                      synthesizing a particular protein

A

Molecules 2

THE PROCESSES OF TRANSCRIPTION AND TRANSLATION

After the DNA molecule partially unwinds and puffs out the surface of the chromosome, a gene in the DNA molecule has been exposed. The codons in the exposed gene now undergo *transcription*, which is "a process of copying." (If the DNA double ladder never partially unwound, how would the cell ever be able to make a copy or transcript of one gene in it?)

The transcription (DNA copying) process creates a new molecule, called *messenger RNA* (abbreviated as *mRNA*). (Review Figure 4.7.) The messenger RNA molecule moves out of the nucleus and onto the surface of a *ribosome* (**RYE**-boh-**sohm**). A ribosome is a tiny, dark "body" (*some*) that contains the sugar, *ribose* (**RYE**-bohs).

The mRNA molecule attaches to the ribosome. Here, the process of *translation* takes place. Essentially, the *nitrogen base language* of the mRNA codons is *translated* or changed into the corresponding *amino acid language* of a particular protein.

Another type of RNA, called *transfer RNA* or *tRNA*, plays a pivotal role in this translation-into-a-protein process. Each individual transfer RNA (tRNA) molecule codes for, and binds to, a certain amino acid within the cytoplasm. During translation, then, a group of tRNA molecules, along with their attached amino acids, lines up across from the mRNA molecule on a ribosome.

The tRNA molecules match their bases up against the complementary bases of the mRNA molecule. The amino acids attached at the other end of the tRNAs link together via *peptide* (**PEP**-tide) *bonds*. The end result is a finished protein or *polypeptide* (**PAH**-lee-**pep**-tide) – a combination of "many" (*poly-*) amino acids connected by peptide bonds in a coded order.

Each completed protein (polypeptide) now detaches from a ribosome and begins to perform its special function within the cell.

Capsulizing transcription and translation, we have:

F

Transcription = A copying process whereby mRNA makes a copy or *transcript* of the bases found within a certain gene in the DNA;

and

Molecules 8

Translation = A "going across" (*trans-*) from the *nitrogen base language* of mRNA into the *amino acid language* of a synthesized protein, with the help of tRNA

The Tragedy of Mutation: Pathophysiology and "Bad" Proteins

Several times we have noted that pathological anatomy, usually associated with a certain degree of Biological Disorder within body structures, frequently results in pathophysiology and morbidity (disease) or mortality (death). In certain cases, this is also true for *mutations* (mew-**TAY**-shuns) – abnormal "changes" in the structure of body proteins.

Consider, for example, the frequently fatal disorder called *sickle cell disease*. In persons with this disease, many of the normally rounded *red blood cells* assume the strange, "sickled" shape of a half moon or crescent. This deforming of the *RBC* (red blood cell) shape, surprisingly enough, is due to a mutation or error in just a *single* amino acid within the long *hemoglobin* (**HE**-moh-**gloh**-bin) molecule.

Hemoglobin is the main oxygen-carrying protein in our red blood cells (RBCs). Due to an error in the gene coding for hemoglobin synthesis, the resulting abnormal hemoglobin molecules in a person with sickle cell disease contain just *one wrong amino acid*! Nevertheless, this seemingly minor occurrence of a Structural Disorder at the Molecular Level (hemoglobin) creates an abnormal hemoglobin that can produce deadly consequences for the patient!

Molecules 1

The sickled (crescent-shaped) RBCs are very fragile, and their plasma membranes extremely sticky. Therefore, there is great risk of suffering a *sickle cell crisis*, wherein many sickled RBCs suddenly clump together or *agglutinate* (ah-**GLUE**-tih-**nayt**) in large numbers. Such large masses of agglutinated (ah-**GLUE**-tih-**nay**-ted) RBCs may completely *occlude* (ah-**KLOOD**) or "stop up" the openings of major blood vessels. As a result, considerable numbers of body cells and tissues may die, due to lack of adequate blood flow and its vital oxygen supply.

Quiz

Refer to the text in this chapter if necessary. A good score is at least 8 correct answers out of these 10 questions. The answers are listed in the back of this book.

1.   The cell organelles, like the entire cell, are:
      (a) Living structures with their own physiology

(b) Tiny organ-like structures having highly specific functions within the cell

(c) Each wrapped in a single plasma membrane

(d) Usually functioning poorly (if at all!)

2. By describing something as "selectively permeable," it means that:
   (a) The object is a totally ineffective barrier
   (b) Neither liquid nor solid can penetrate it
   (c) Some substances pass through it, while the movement of others is prevented
   (d) Energy is always required to cross through it

3. The Principle of "like dissolves like" implies that:
   (a) Water and oil don't mix
   (b) The phospholipid bilayer is easily penetrated by Na^+ ions
   (c) Water and oil sometimes can mix well together
   (d) Charged, polar particles react well with those that are un-charged

4. A condition of thrusting of H_2O molecules:
   (a) Osmosis
   (b) Active transport
   (c) Facilitated diffusion
   (d) Mediated transport

5. Glucose enters most body cells by what specific process?
   (a) Active transport
   (b) Mitosis
   (c) Occurrence of a diffusion equilibrium
   (d) Facilitated diffusion

6. A sodium–potassium exchange pump is a good example of:
   (a) Cytokinesis
   (b) Diffusion down a concentration gradient
   (c) Facilitated diffusion through pores
   (d) Energy-primed movement up a concentration gradient

7. The ADP molecule becomes much more suited for supplying cell energy when:
   (a) Mutations render it into a less harmful conformation
   (b) A terminal phosphate group breaks off the end of it
   (c) Nitrogen atoms are fused to it in great numbers
   (d) A terminal phosphate group is added

8. Literally means the "process of breaking down sweets or glucose":
    (a) Glycolysis
    (b) Glycogen
    (c) Krebs cycle
    (d) Amino acid transference

9. Cellular respiration occurs within the _____ of the mitochondrion:
    (a) Outer membrane
    (b) Ribosomes
    (c) Cristae and matrix
    (d) Matrix only

10. DNA codons function primarily as:
    (a) Alternative functions for membrane lipids
    (b) Protein carriers
    (c) Code words for amino acids
    (d) Translation of the language of peptide bonds into the language of ionic bonds

Body-Level Grids for Chapter 4

Several key body facts were tagged with numbered icons in the page margins of this chapter. Write a brief summary of each of these facts into a numbered cell or box within the appropriate *Body-Level Grid* that appears below.

Anatomy and *Biological Order* **Fact Grids for Chapter 4:**

A

MOLECULE
Level

1	2

ORGANELLE
Level

1

Function and Biological Order **Fact Grids for Chapter 4:**

MOLECULE
Level

1	2

3	4

5	6

7	8

ORGANELLE
Level

1	2

Anatomy and *Biological Disorder* Fact Grids for Chapter 4:

A

MOLECULE
Level

1

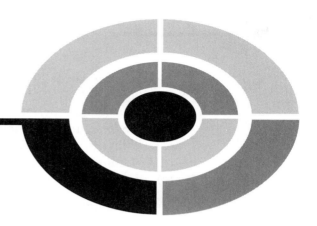

Test: Part 2

DO NOT REFER TO THE TEXT WHEN TAKING THIS TEST. A good score is at least 18 (out of 25 questions) correct. Answers are in the back of the book. It's best to have a friend check your score the first time, so you won't memorize the answers if you want to take the test again.

1. The Chemical Level of Body Organization consists of:
   (a) Subatomic particles, organelles, and molecules
   (b) Cells, atoms, and molecules
   (c) Subatomic particles, atoms, and molecules
   (d) Atoms, tissues, and elements
   (e) Practically everything within the Great Body Pyramid!

2. Anatomy joins with the Atomic Theory in explaining that:
   (a) All body structures consist of atoms and the empty spaces between these atoms
   (b) Several levels of human anatomy are nothing but atoms!
   (c) Atoms, like anatomy, have a corresponding physiology
   (d) Anatomy, like atoms, consists of structures that are "uncuttable"
   (e) Neither area of human study involves any practical use of dissection

3. Nonpolar covalent bonding differs from polar covalent bonding in its:
   (a) Transfer of innermost electrons between atoms
   (b) Tendency to break down into ions
   (c) Ability to share electron clouds equally between atoms
   (d) Frictional resistance to magnetic forces
   (e) Trend toward tearing apart into teeny little pieces

4. Very common cation within the body's extracellular fluid:
   (a) Cl^-
   (b) U-235
   (c) Na^+
   (d) K^+
   (e) HPO_4^-

5. A log sitting on the ground has much more of this energy, compared to a log burning in a fire:
   (a) Anabolic
   (b) Metabolic
   (c) Kinetic
   (d) Human
   (e) Potential

6. Synthesis reactions are alternatively known as:
   (a) Protein catalysts
   (b) Anabolism
   (c) Destructogenesis
   (d) Catabolism
   (e) Exchange reactions

7. The so-called intestinal neutralization equation helps to "neutralize":
   (a) $NaHCO_3$
   (b) NaCl
   (c) H_2O
   (d) H_2CO_3
   (e) HCl

8. Serve as important H^+ ion acceptors:
   (a) Sponges
   (b) Acids
   (c) Bases or alkali
   (d) Electrolytes
   (e) Most cations

9. The pH of a neutral solution is approximately:
   (a) 3
   (b) 5
   (c) 7
   (d) 7.4
   (e) 14

10. A bottle containing household ammonia has:
    (a) An extremely high pH
    (b) A low pH but a high H^+ ion concentration
    (c) Most of the same acid–base characteristics as does pure water
    (d) Both a high pH and a high H^+ concentration
    (e) Both a low pH and a low H^+ ion concentration

11. When Baby Heinie falls into a vat of lemon juice (containing lots of ascorbic acid) and gulps down large quantities of it, he is in most danger of suffering from:
    (a) A gall bladder attack!
    (b) Chronic acne
    (c) Metabolic acidosis
    (d) Respiratory alkalosis
    (e) Metabolic alkalosis

12. The cell is considered the lowest level where microphysiology exists, because:
    (a) The cell organelles are considered the "life-line"
    (b) Only groups of related cells are actually functioning
    (c) Cell structure determines cell function
    (d) True physiology starts with the basic unit of life
    (e) Anatomy starts at the cell level

13. The portion of the cell or plasma membrane most resembling a soap bubble:
    (a) Polar phosphate-nitrogen head groups of the phospholipids
    (b) Water-filled channels through transport proteins
    (c) Nonpolar fatty acid tails of the phosopholipids
    (d) Sodium chloride electrolyte content
    (e) Bubbly bath of acid residues

14. Open a bottle of perfume. Soon, persons in the far corners of the room can smell it, due mainly to the process of:
    (a) Cytokinesis
    (b) Mediated transport

(c) Active transport
(d) Chromosome breakage
(e) Simple diffusion

15. A newly planted flower begins to droop and wilt. You water it, and it springs right back, owing to the pressure created by:
(a) Net osmosis into the plant cells
(b) Uncontrolled drying out in the baking sun!
(c) Simple diffusion of electrolytes out of the leaves
(d) Net osmosis out of the plant cells
(e) The occurrence of a permanent diffusion equilibrium

16. If a special drug were given that destroyed all carrier proteins in the cell membrane:
(a) Simple diffusion through the membrane would grind to a halt
(b) No H_2O could fit through membrane channels
(c) All mediated transport processes would stop
(d) Particles would still tend to move randomly uphill across the membrane
(e) The cell membrane would completely fall apart!

17. Enters most body cells by a facilitated diffusion process:
(a) Bases and alkali
(b) Na^+ and K^+ ions
(c) Clusters of H_2O molecules
(d) Fatty acid
(e) Glucose

18. Even though sodium ions continually diffuse into the cell through channels, the inside of the cell remains negatively charged compared to the outside, because:
(a) Too many potassium ions are around!
(b) Na^+ ions are actively transported back out of the cell at a faster rate
(c) The sodium is broken down into its individual subatomic particles
(d) The cell produces lots of Na^+ during its aerobic metabolism
(e) K^+ ions are pumped back into the cell using energy from ATP

19. Excess free energy from catabolism of cell nutrients is usually transferred to:
(a) Adenosine diphosphate
(b) ATP

    (c) The outer mitochondrial membrane

    (d) $NaHCO_3$

    (e) Potential energy that is seldom, if ever, tapped

20. Two pyruvic acids result from this sequence of chemical reactions:
    (a) Krebs cycle
    (b) Gene mutation
    (c) Glycolysis
    (d) Mediated transport
    (e) Osmosis

21. The fluid matter of the cell:
    (a) Mitochondria
    (b) Cristae
    (c) Microtubules
    (d) Cytoplasm
    (e) Nucleolus

22. A site for the electron transport system:
    (a) Outer mitochondrial membrane
    (b) Mitochondrial matrix
    (c) "Crests" of the inner mitochondrial membrane
    (d) Cytoplasm
    (e) Plasma membrane

23. Chemical serving as the final acceptor for "energy-depleted" electrons in the electron transport system:
    (a) C
    (b) N
    (c) H
    (d) O
    (e) P

24. Total number of ATP molecules produced per glucose molecule catabolized under aerobic conditions:
    (a) 16
    (b) 22
    (c) 32
    (d) 34
    (e) 38

25. The phase of protein synthesis where a copy of the DNA codons is constructed as a series of corresponding codons in mRNA:
   (a) Interphase
   (b) Transcription
   (c) DNA synthesis
   (d) Catabolism
   (e) Translation

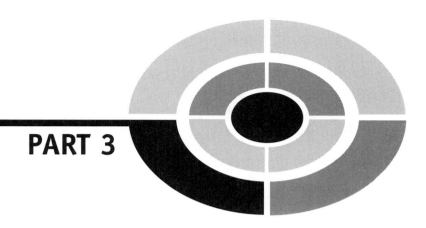

PART 3

The Physiology of Skin, Bone, and Muscle

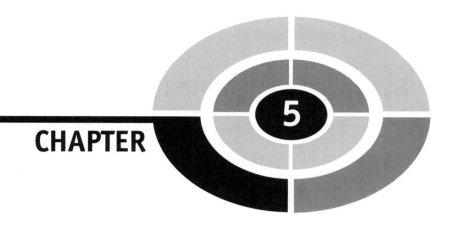

Working the Human "Hard" Drives: The Physiology of Skins and Skeletons

So far, this book has been easy. But now it is time to get "hard"! Specifically, it is time for us to start working the human "hard" drives – the physiology of our tough outer skin and our rock-like, "inner" (*endo*-) "hard dried body," the *endoskeleton* (**EN**-doh-**skel**-uh-ton).

"Systemic" Physiology of the Skin

For our purposes in this book, we will concentrate upon several aspects of the "systemic" physiology of the skin. By this we mean the broad, organ system and organism effects of skin physiology, rather than the mostly "local" physiological effects, such as protection from the *ultraviolet* (**ul-**truh-**VEYE**-uh-lit) or *UV rays* in sunlight.

SYSTEMIC PHYSIOLOGY 1: PROTECTING THE BODY FROM DEHYDRATION

Organ 1

The skin, and especially the epidermis, provides the entire human organism absolutely vital protection from massive tissue *dehydration* (**dee**-high-**DRAY**-shun). This term literally means "the process of" (*-tion*) losing "water" (*hydr*) "from" (*de-*) the body tissues.

You may recall (Chapter 3) that water or H_2O is the major polar covalent molecule within the body fluids. As such, it is the chief solvent ("dissolver") of sodium chloride (NaCl) and other charged *solutes* (**SAHL**-yoots) – "things dissolved." Hence, our "internal sea" of body fluids is principally an *aqueous* (**AH**-kwee-us) or "watery" *solution* (**so-LOO**-shun). A solution, in general, is a liquid containing dissolved substances (one or more types of solute particles). Summarizing yields:

$$\text{"INTERNAL SEA" OF BODY FLUIDS} = \underset{\textit{("Things Dissolved")}}{\text{CHARGED } SOLUTES} + \underset{\textit{("The Dissolver")}}{\text{WATER } SOLVENT} = \text{AN AQUEOUS } SOLUTION$$

The aqueous solution of our internal sea, the astute reader may remember (Chapter 2), is called the extracellular fluid or *ECF* – the fluid lying "outside of the little cells" in our body tissues. The ECF also includes the *blood plasma* (**PLAZ**-muh), or watery fluid "matter" of the blood. These ECF and blood plasma concepts further bring back to mind the intracellular fluid or *ICF* "within" (*intra-*) our cells. The intracellular fluid (ICF), like the extracellular fluid (ECF), is essentially a water-dominated aqueous solution containing various dissolved solutes.

Normal hydration and osmotic equilibrium

When the body is in a state of normal *hydration* (high-**DRAY**-shun), it is normally "watered" (*hydr*). There is a chemical equilibrium, or dynamic balance, between the water concentrations of the ECF (including the blood plasma) and the ICF. (Consult Figure 5.1, A.)

Back in Chapter 4, we talked about osmosis. This is the "thrusting" (*osm*) of H_2O molecules in a particular direction as they diffuse across the cell's plasma membrane – either from the ECF into the ICF, or from the ICF out into the ECF.

If the water concentration of the ECF equals the water concentration of the ICF, there is no water concentration gradient between them. Thousands of H_2O molecules diffuse back and forth through the water-filled pores of the selectively permeable cell membrane, entering and leaving the cell at equal rates. Thus, a condition of *osmotic* (ahz-**MAHT**-ik) *equilibrium* exists.

Isotonic body fluids and stable cell size

Pretend that we are looking at the *red blood cells* (*RBCs*) within the bloodstream of a normally *hydrated* (**HIGH**-dray-ted) person (Figure 5.1, B). In such a person, their blood plasma is said to represent an *isotonic* (**eye**-so-**TAHN**-ik) *solution*. An isotonic solution is a solution whose "strength" or "concentration" (*ton*) of solute is the "same" (*iso-*) as that found within the intracellular fluid. Now, the ICF of red blood cells is an aqueous solution with a strength or concentration of about 0.9% NaCl solute dissolved in 99.1% H_2O solvent. If the surrounding blood plasma has the same concentration of solute and water as this, then it is classified as an isotonic solution. This classification as isotonic is important, because it results in stability of cell size. Specifically, there is no concentration gradient for H_2O across the cell membrane of the RBC, so that there is an osmotic equilibrium. Water molecules diffuse into the RBCs, just about as fast as they diffuse out. When RBCs travel through very narrow blood *capillaries* (**CAP**-uh-**lair**-eez), for instance, not being too big allows them to easily pass (without getting stuck!). To summarize,

NORMAL TISSUE HYDRATION: (A Normal Volume of Body Water)	$\longrightarrow$	BLOOD PLASMA AND OTHER PARTS OF ECF ARE *ISOTONIC SOLUTIONS*
		$\downarrow$
STABILITY OF RBC SIZE IS ACHIEVED	$\longleftarrow$	*NO NET OSMOSIS* OF H_2O EITHER INTO OR OUT OF CELLS

Tissue dehydration results in cell shrinkage

Following from our consideration of normal hydration and the ECF as an isotonic solution, we can now see how this delicate balance is upset during tissue dehydration. The fairly watertight skin normally prevents a huge volume of our body water content from simply evaporating into the air from our watery tissues. This is largely due to the presence of *keratin* (**CARE**-uh-**tin**) *granules* scattered throughout the surface epidermis. Keratin is a family of tough, waterproof proteins founds in the "horns" (*kerat*) of animals, as well as in the skin, nails, and hair of humans.

"But what if a large portion of my skin (along with its waterproof keratin protein) is gone, due to a severe burn?" Most unfortunately, besides a huge risk of severe bacterial infection, your soft, wet, underlying connective tissues and muscle will tend to lose huge amounts of H_2O to the surrounding air by evaporation. This will soon result in severe tissue dehydration. And in the blood plasma, you are quite likely to suffer extreme *hemoconcentration* (**HE**-moh-kahn-sen-**TRAY**-shun). This is a dramatic rise in the "blood" (*hemo*) solute "concentration," far above the normal 0.9% NaCl figure.

With so little water remaining, the blood plasma (and other portions of the ECF) become a *hypertonic* (**high**-per-**TAHN**-ik) *solution*. This solution type has an "excessive or above normal" (*hyper-*) "strength or concentration" (*ton*) of solute. As displayed in Figure 5.1 (C), a hypertonic solution, because it has such a high solute concentration, has a very low solvent (H_2O) concentration. Hence, the water concentration in the ICF of the red blood cells is greater than the water concentration in the surrounding blood plasma. There is a net osmosis of thousands of H_2O molecules out of the RBCs and into the overly concentrated plasma. *Crenation* (kree-**NAY**-shun) – the "process of" (*-tion*) becoming "notched" (*cren*) – happens to the RBCs.

The abnormally shrunken, *crenated* (**KREE**-nay-ted) red blood cells, bearing numerous wrinkles and notches, can no longer function normally in carrying oxygen (O_2) molecules through the bloodstream. And the dehydrated person may also suffer from severe tissue *ischemia* (is-**KEY**-me-ah), a "holding back" of blood flow that deprives many parts of the body.

In summary,

Cell 1

SEVERE TISSUE DEHYDRATION: (Large Amount of Body Water Loss) $\longrightarrow$ BLOOD PLASMA AND OTHER PARTS OF ECF BECOME *HYPERTONIC SOLUTIONS*

$\downarrow$

CRENATION (SHRINKING AND NOTCHING) OF RBCs $\longleftarrow$ NET OSMOSIS OF H_2O *OUT OF* RBCs

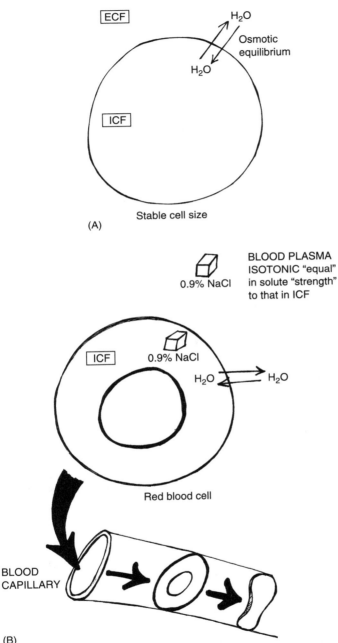

Fig. 5.1 Effects of body water content on osmosis and RBCs. (A) Normal tissue hydration and osmotic equilibrium. (B) Isotonic body fluids and stable cell size.

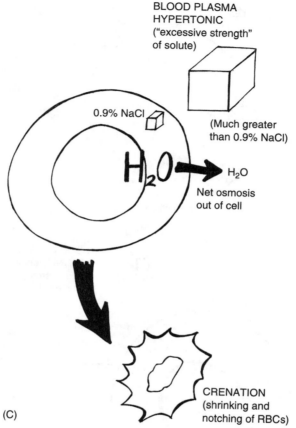

(C)

Fig. 5.1   (continued) (C) Hypertonic body fluids and cell shrinkage.

SYSTEMIC PHYSIOLOGY 2: PROVIDING THE BODY WITH THERMOREGULATION

In addition to helping protect the body from dehydration, the skin also has a systemic influence upon thermoregulation. Chapter 2 defined this as the regulation of body heat or temperature. Go back and review the text accompanying Figure 2.9.

You will note several components of the skin operating in the negative feedback control system that helps maintain homeostasis of oral body temperature in degrees Fahrenheit. These are:

(1) Skin thermoreceptors in the dermis detect a rise or fall in oral body temperature from its set-point level.

(2) Sweat glands in the dermis release sweat into a sweat duct, which travels through the epidermis and opens onto the skin surface, helping to cool the body by evaporation.

(3) Vasodilation (widening) of dermal blood vessels helps reduce oral body temperature by the process of heat radiation.

Our Endoskeleton: Bones and Their Tissues

Move in deeply enough through our tough skin, and you are bound to hit something even harder – a *bone organ* of our endoskeleton. A number of specific bones and types of bone shapes are talked about within the pages of *ANATOMY DEMYSTIFIED*. It is our task in this physiologically-oriented book, however, to focus upon what these bone organs *do*, systemically, for the body as a whole.

INTRODUCTION TO DENSE OR COMPACT BONE TISSUE

An important part of every bone organ is its hard, white, *dense or compact bone tissue*. As Figure 5.2 (A) reveals, the dense (compact) bone tissue is mainly found as the thin, white, outer shell of calcium-rich *matrix* (**MAY**-tricks). It is described as dense or compact because, to your unaided eyes, the bone matrix appears to be uniformly white, uninterrupted, and solid as a rock!

A very different reality emerges, however, when a thin slice of the dense bone tissue is stained and viewed through a microscope (Figure 5.2, B). From this detailed perspective, the broad field of white bone matrix is seen to be shot through with many large black holes. (This makes it look more like Swiss cheese than white rock, doesn't it?)

Each black hole is technically called a *central or Haversian* (hah-**VER**-shun) *canal*. The dark hole is called a *central* canal because it resembles a black bull's-eye in the *center* of a target. Its strange name, Haversian, literally "refers to" (*-ian*) Clopton *Havers* (**HAH**-vers). He was an English physician and anatomist (living from 1650 to 1702) who first described much of bone microanatomy.

Arranged around the central (Haversian) canal is a series of rings of bone matrix, of progressively increasing diameter. The outer edge of each ring is framed by a series of small, dark pits called *lacunae* (lah-**KOO**-nee) or "lakes." The entire circular pattern around each central canal is called

A

Tissue 1

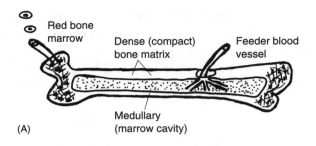

(A)

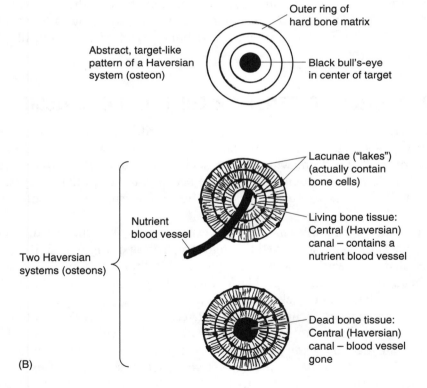

(B)

Fig. 5.2   The gross and microscopic anatomy of bone tissue. (A) Gross anatomy of a long bone (including bone matrix). (B) Microanatomy of dense bone matrix: a series of Haversian systems.

a *Haversian system* or *osteon* (**AHS**-tee-**ahn**). The Haversian system (osteon) is the major repeating anatomical subunit that makes up dense bone tissue.

The "calcium-connection": osteoblasts create the bone matrix

The white, rock-hard bone matrix makes up the majority of each Haversian system or osteon. "Why is the bone matrix *white*, Professor?," the bone-enthused reader may ask of Joe, the Talking Skeleton. Being a scholar, Professor Joe will hesitate to tell us, immediately. Rather, he wants us to think about the reason as a part of *ossification* (**ah**-sih-fih-**KAY**-shun) – the "process of bone formation."

In the human *embryo* or "sweller," the long bones of the skeleton are principally composed of *cartilage connective tissue*, rather than bone. As Figure 5.3 shows, the original femur, for instance, is actually a miniature *cartilage model* of the final adult bone. Cartilage (more commonly known as "gristle") is a soft, rubbery connective tissue that is *avascular* (a-**VAS**-kyoo-lar). This means that cartilage does "not" (*a-*) contain any "little vessels" (*vascul*) for carrying blood.

But as development of the embryo continues, blood vessels invade the cartilage model. Coming along with them are the *osteoblasts* (**AHS**-tee-oh-**blasts**) or "bone" (*oste*) "formers" (*blasts*). The osteoblasts are large, spider-shaped cells that produce bone collagen fibers. Numerous criss-crossing bone collagen fibers eventually create the basic grid-like anatomy of the bone matrix.

After the collagen gridwork of the bone matrix is laid down, the osteo-blasts extract *calcium* (*Ca^{++}*) *ions*, phosphorus, and other chemicals from the bloodstream. Ossification (bone formation) begins as the osteoblasts deposit sharp, needle-shaped crystals of *calcium phosphate* (**FAHS**-fate) onto the surfaces of the collagen grid. Being snow-white, the calcium phosphate crystals eventually hide the underlying collagen fibers. [**Study suggestion:** Think of each bone collagen fiber as being a long, straight, pretzel stick. In the factory assembly line, the pretzel sticks are covered with white salt (NaCl) crystals as they go through a salting machine. Now, suppose that a pretzel stick gets jammed in the salting machine overnight. The next morning, the salter is finally stopped, and the pretzel stick (collagen fiber) is at last removed. Will you be able to see the pretzel stick? Why, or why not? How is this like the situation in the bone matrix?]

Cell 2

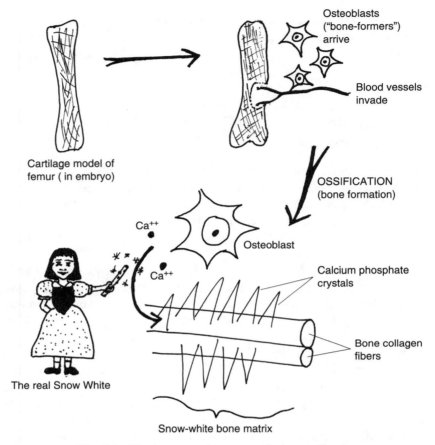

Fig. 5.3 The bone matrix appears during ossification.

By the time the child becomes an adult, her femur is mostly snow-white. You might, therefore, consider the bone matrix as essentially being a storage bank for calcium ions.

"Systemic" Physiology of the Bones

Now that we know something about bone microanatomy, we are ready to go forward and examine several examples of the "systemic" physiology of bones.

SYSTEMIC BONE PHYSIOLOGY 1: MAINTAINING HOMEOSTASIS OF BLOOD [Ca^{++}]

As a result of ossification, millions of calcium (Ca^{++}) ions become stored in the form of calcium phosphate crystals on the bone collagen fibers. Even though the bones eventually become fully ossified, however, there is a process of *bone remodeling* that continues throughout life.

Bone remodeling is a change in the shape, thickness, and strength of bones that occurs under the influence of various factors. Important among these factors is the blood calcium ion concentration, which can be denoted with brackets as *blood [Ca^{++}]*.

P

Organ 2

Blood calcium ion concentration, or blood [Ca^{++}], is an anatomical parameter that exists at the Chemical Level of Body Organization. We can clearly visualize each calcium ion as a tiny black dot with two positive (+) charges, circulating within the bloodstream (Figure 5.4).

The blood [Ca^{++}] is usually measured in units of *mg/dL* – milligrams of calcium ions per *deciliter* (**DEH**-sih-**lee**-ter) – of blood. A deciliter is one "tenth" (*deci-*) of a "liter."

The critical balance of normocalcemia

You may recall some important general concepts from Chapter 2, such as the ideas of the normal range of a body parameter, its set-point value, its lower normal limit (LNL), and its upper normal limit (UNL). We have already outlined these concepts for the physiological parameter of oral body temperature in degrees Fahrenheit.

Taking a similar approach, we can state that the normal or *reference range* for blood [Ca^{++}] is from a lower normal limit of 8.5 mg/dL blood up to an upper normal limit of 10.6 mg/dL blood. Using the same naming model as for normothermia, we will state this condition as *normocalcemia* (**nor**-moh-kal-**SEE**-me-uh). Normocalcemia exactly translates to mean "a normal" (*normo-*) "blood condition of" (*-emia*) "calcium" (*calc*).

Assuming that normocalcemia is maintained over a definite period of time, we can further say that there is a homeostasis or relative constancy of the blood calcium ion concentration (Figure 5.4).

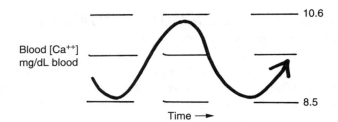

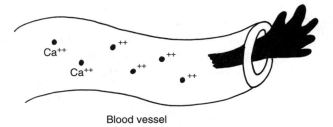

Fig. 5.4. Normocalcemia and homeostasis of blood calcium concentration.

Blood calcium concentration: avoiding too much, or too little

"So, what's the big deal about maintaining homeostasis of blood calcium ion concentration, Professor Joe?," the untutored mind may be prodded to ask. We know from our earlier studies, as for body temperature (Chapter 2), that too much, or too little, of a good thing (particular anatomical or physiological parameter) is a form of Biological Disorder. In the case of body temperature, a rise beyond the upper normal limit leads to a particular hyper- (excessive or above-normal) state, called hyperthermia. For blood [Ca^{++}], the appropriate term would be *hypercalcemia* (**high**-per-kal-**SEE**-me-uh) – "a blood condition of" (*-emia*) "above normal or excessive" (*hyper-*) "calcium" (*calc*). Hypercalcemia essentially represents a failure of homeostasis of blood calcium ion concentration that occurs at the upper end of the normal range (Figure 5.5, A). The blood vessels become over-packed with Ca^{++} ions.

At the opposite end of the calcium concentration range, a fall below the lower normal limit results in a particular hypo- (below normal or deficient) state. This deficient state is technically labeled as *hypocalcemia* (**high**-poh-kal-**SEE**-me-uh). In particular, hypocalcemia is extremely dangerous to human

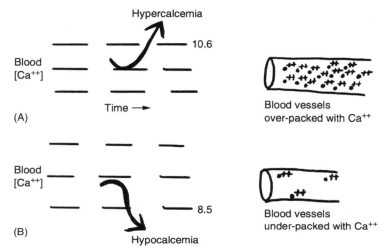

Fig. 5.5   Hypercalcemia and hypocalcemia: failures of homeostasis of blood calcium concentration.

well-being and survival – Why is this? The reason is that an adequate amount of calcium ions within the bloodstream is absolutely necessary for all of the muscle tissue in the body to contract. Hypocalcemia, therefore, can lead to muscular *hypoexcitability* (**high**-poh-eks-sight-ah-**BILL**-ih-tee). If the cardiac muscle tissue in the heart wall is "deficiently excited," then watch out for heart failure! And if the skeletal muscle tissue attached to our bones is "deficiently excited," then you literally may not be able to "move a muscle"!

Tissue 1

Thinking in terms of homeostasis, we have a failure of relative constancy of blood [Ca^{++}] and the introduction of Biological Disorder at the lower end of the normal range (Figure 5.5, B).

Bone matrix and negative feedback control of blood [Ca^{++}]

From the above discussion, maintaining homeostasis of blood [Ca^{++}] is important, but avoiding hypocalcemia is critical for survival!

Our body utilizes the bone matrix, with its heavy concentration of Ca^{++} ions deposited as crystals of calcium phosphate, as a type of blood calcium bank. After severe exercise, say, or maybe a day of living without consuming any calcium in the diet, the blood [Ca^{++}] may decline toward its lower normal limit. This drop in blood calcium level serves as a stimulus, which engages a negative feedback control system (see Figure 5.6.)

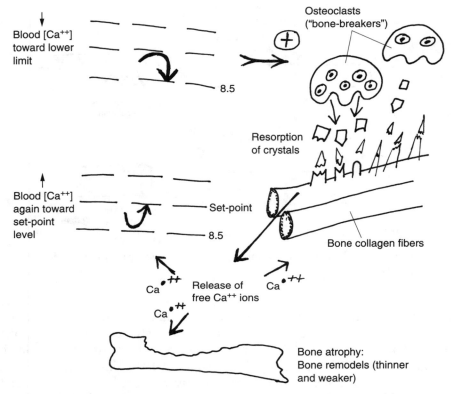

Fig. 5.6   A negative-feedback response to a drop in blood [Ca^{++}].

This stimulus is especially strong for the *osteoclasts* (**AHS**-tee-oh-**klasts**) or "bone-breakers" (-*clasts*). The osteoclasts are definitely not enforcers working for the Mafia, who will "break your bones" if you don't pay up! In reality, the osteoclasts are large, *multinucleate* (**mul**-tee-**NOO**-klee-ut) cells that release special digestive enzymes. [**Study suggestion:** Look at Figure 5.6. See how far you can translate the word multinucleate into its common English equivalent.]

The digestive enzymes from the osteoclasts cause a partial *resorption* (rih-**SORP**-shun) or "drinking-in again" of the bone matrix. This occurs because some of the calcium phosphate crystals on the bone collagen fibers are broken down and dissolved. This releases free Ca^{++} ions back into the bloodstream. As a result, the blood calcium ion concentration rises back toward its set-point (long-term average) value. By this calcium-withdrawal process from our bone matrix "bank," then, hypocalcemia is usually prevented. As a consequence of the resorption process, however, the bones remodel and become thinner and weaker. If the resorption process goes on long enough, as in a person whose diet is chronically low in calcium, a condition of *bone atrophy* (**AT**-roh-fee) may result. Atrophy is literally "a condition" (*-y*) "without" (*a-*) "nourishment" (*troph*). In bone atrophy, the bones are not being "nourished," in the sense that they are not being stimulated to add more calcium phosphate crystals. If the atrophy is severe enough, the bones are also much more vulnerable to deformity and fracture.

By reversing our previous thinking, we can trace the response of the body to a rise in the blood calcium ion concentration toward its upper normal limit (Figure 5.7). This rise is a stimulus for the osteoblasts (bone-makers). As

Organ 1

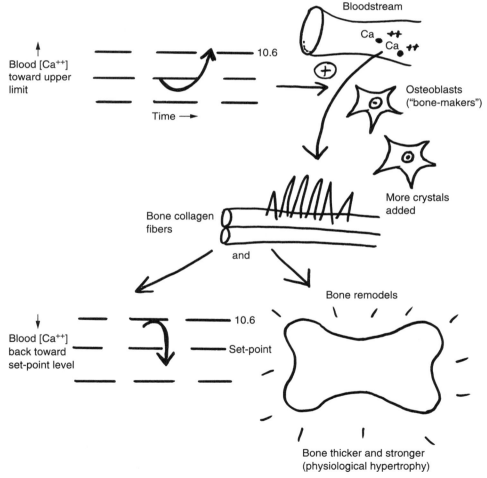

Fig. 5.7   A negative feedback response to a rise in blood [Ca^{++}].

they do during the process of ossification, the osteoblasts extract free Ca^{++} ions from the bloodstream, and supervise their deposition upon bone collagen fibers as more calcium phosphate crystals.

As a result, the blood $[Ca^{++}]$ falls back down toward its set-point level, and the abnormal condition of hypercalcemia is usually prevented. As a secondary consequence of "depositing" more calcium "money" to the bone matrix "bank," bone remodeling occurs. But this time it is in the direction of *bone hypertrophy* (high-**PER**-troh-fee) – "a condition of" (-*y*) "excessive or above normal" (*hyper-*) "nourishment" (*troph*) or stimulation. In short, the osteoblasts withdraw more calcium from the bloodstream, and deposit it onto bone collagen fibers as more calcium phosphate crystals. With more crystals, the bone *hypertrophies* (high-**PER**-troh-fees), becoming both thicker and stronger. This hypertrophy (excessive "nourishment" or stimulation) is completely normal and natural. Therefore, sometimes it is called *physiological hypertrophy*. This is an increase in the size of some body structure that is due to normal, rather than pathological, changes (such as a bone tumor).

Organ 3

SYSTEMIC BONE PHYSIOLOGY 2: THE ROLE OF RED MARROW IN HEMATOPOIESIS

Let us return to the gross anatomy of a long bone. This time, though, we are going to cut through the entire bone, in a *longitudinal* ("lengthwise") direction. As Figure 5.8 reveals, there is another type of bone tissue present, in addition to dense (compact) bone tissue. This second type is called *spongy or cancellous* (**CAN**-sih-**lus**) *bone tissue*. This type of bone tissue resembles a sponge, since its hard white matrix is arranged into criss-crossing *trabeculae* (trah-**BEK**-yuh-**lie**) or "little beams." These trabeculae (little beams) of matrix, since they extensively criss-cross, create a cancellous pattern, which literally "pertains to" (-*ous*) a "lattice" (*cancell*).

A sponge, of course, is filled with holes. In spongy or cancellous bone, the holes contain the *red bone marrow*. The red bone marrow largely consists of a soft, pulpy meshwork of reddish-colored blood vessels. These blood vessels snake in and out of the holes within the latticework of slender white trabeculae of spongy bone.

Tissues 1

The major function performed by red bone marrow is *hematopoiesis* (**he**-muh-toh-poy-**EE**-sis) – the process of "blood" (*hemat*) "formation" (-*poiesis*). By this translation, however, we do not mean that all of the components of the blood are manufactured by the red bone marrow. Rather, we mean the so-called *formed elements* of the bloodstream are manufactured here. In particular, these are the various types of blood cells and blood cell fragments, which are the "elements" of the blood having a specific shape or "form."

Hemocytoblasts: the original parent cells

Cell 3

Any close, microscopic look at the red bone marrow (Figure 5.8, B), shows some of the *stem* or *parent cells*, which are the predecessor cells eventually giving rise to the formed elements. Everything begins with the *hemocytoblasts* (**he**-moh-**SIGH**-toh-blasts), literally the "blood" (*hem*) "cell" (*cyt*) "formers" (*-blasts*). Each hemocytoblast is a fairly large cell having a distinct oval nucleus and projecting arms of cytoplasm. Through a number of intermediate cell stages, the hemocytoblasts eventually create all the formed elements found in the bloodstream.

Let us first consider the *erythrocytes* (eh-**RITH**-roh-**sights**) or "red" (*erythr*) "cells." Each mature erythrocyte is an *anucleate* (ay-**NEW**-klee-ut), *biconcave* (**BUY**-kahn-**cave**) *disc*. This means that the erythrocyte or RBC is shaped somewhat like a red hourglass (when viewed on its side), being "caved in" (*concave*) on "both" (*bi-*) sides.

One of its intermediate stem cells is the *erythroblast* (eh-**RITH**-roh-**blast**) or "red-former." The erythroblast is a regular cell with a nucleus, but it loses this nucleus on its way to becoming a mature erythrocyte. [**Study suggestion:** What does anucleate mean?] Of course, the erythrocyte looks red because it can contain up to 280 million red-colored *hemoglobin* (**HE**-moh-**gloh**-bin) molecules, which do its main job of carrying oxygen through the bloodstream.

The *leukocytes* (**LOO**-koh-**sights**) or "white" (*leuk*) cells are named for the clear, whitish color of their cytoplasm. Unlike the erythrocyte, all leukocytes are regular nucleated cells. The most common type of leukocyte created during hematopoiesis is the *neutrophil* (**NEW**-troh-**fil**). The neutrophil gets its name from its apparent "love or fondness" (*phil*) for "neutral" (*neutr*) dyes during biological staining. Just after the hemocytoblast, the neutrophil's parent cell is the *myeloblast* (my-**EL**-oh-**blast**) or "former" (*blast*) in the bone "marrow" (*myel*).

Finally, we come to the blood *platelets* (**PLAY**-tuh-lets) or *thrombocytes* (**THRAHM**-boh-**sights**). The platelets are literally "small plates" – small, plate-like fragments of disintegrated bone marrow cells. Just after the hemocytoblast, the platelet's parent cell is the *megakaryoblast* (**meg**-uh-**CARE**-ee-oh-blast) or "large" (*mega-*) "nucleus" (*karyo*) "former" (*blast*). The megarkaryoblast is an immature cell that eventually becomes a mature bone marrow cell that breaks apart into a number of platelets. The alternative name

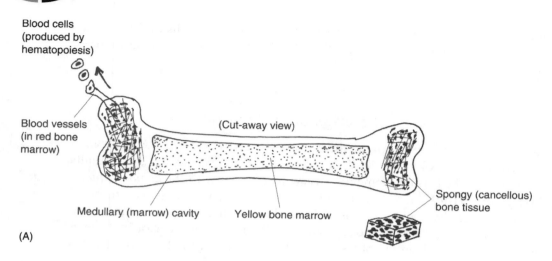

Blood cells
(produced by
hematopoiesis)

Blood vessels
(in red bone
marrow)

(Cut-away view)

Spongy (cancellous)
bone tissue

Medullary (marrow) cavity   Yellow bone marrow

(A)

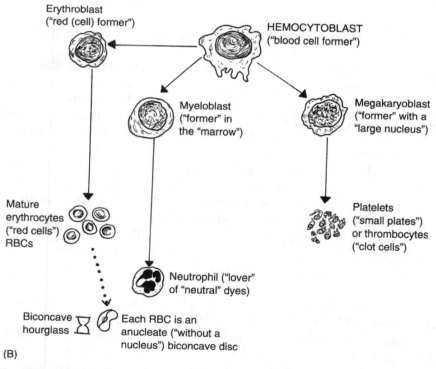

Erythroblast
("red (cell) former")

HEMOCYTOBLAST
("blood cell former")

Myeloblast
("former" in
the "marrow")

Megakaryoblast
("former" with a
"large nucleus")

Mature
erythrocytes
("red cells")
RBCs

Neutrophil ("lover"
of "neutral" dyes)

Platelets
("small plates")
or thrombocytes
("clot cells")

Biconcave
hourglass

Each RBC is an
anucleate ("without a
nucleus") biconcave disc

(B)

Fig. 5.8   Red bone marrow and hematopoiesis of the formed elements. (A) Red bone marrow and location of hematopoiesis within spongy (cancellous) bone tissue. (B) The hemocytoblast and major types of formed elements that are derived from it.

for the platelets is thrombocytes, which means "clot" (*thromb*) "cells" (*cytes*). They get this name from their essential function of sticking together to help create *thrombi* (**THRAHM**-buy) or blood "clots."

Quiz

Refer to the text in this chapter if necessary. A good score is at least 8 correct answers out of these 10 questions. The answers are listed in the back of this book.

1. Layer of the skin containing keratin granules that provide waterproofing:
   (a) Epidermis
   (b) Subdermis
   (c) Accessory layer
   (d) Dermis

2. Charged solutes + water solvent = _____:
   (a) Polar ions
   (b) Dehydration
   (c) Aqueous solution
   (d) Isotonic "dissolver"

3. When an osmotic equilibrium exists across the plasma membrane:
   (a) There is not net diffusion of H_2O either into, or out of, the cell
   (b) A water concentration gradient is present
   (c) Water and dissolved solute are pumped uphill using ATP energy
   (d) H_2O is totally blocked from moving through membrane pores

4. RBCs normally stay small enough to pass through tiny blood capillaries:
   (a) Since blood plasma is usually a hypotonic solution
   (b) Except when the person forgets to urinate!
   (c) Until they explode on their way to the tissues
   (d) Because the ECF is an isotonic solution

5. Hemoconcentration resulting from severe dehydration creates:
   (a) A dramatic drop in the blood plasma NaCl concentration
   (b) Crenation of many circulating erythrocytes
   (c) An abnormally large blood $[H_2O]$
   (d) A huge swelling under the skin

6. The skin's contribution to thermoregulation is chiefly due to the:
    (a) Positive feedback effect of a rise in body temperature upon oral body temperature
    (b) Negative feedback loops involving increased sweating and vaso-dilation
    (c) Removal of all external heat-generating influences
    (d) Attainment of a calm, steady state involving no physical exer-tion

7. A cartilage model in the embryo is often one of the first steps in:
    (a) Ossification
    (b) Dermabrasion
    (c) Leukocytosis
    (d) Osmotic pressure production

8. Osteoblasts deposit these on bone collagen fibers:
    (a) Osteons or Haversian systems
    (b) Pickled turnips
    (c) Cubic NaCl crystals
    (d) Slender needles of calcium phosphate material

9. Hypocalcemia is extremely dangerous to the body, considering that it:
    (a) Can result in total paralysis of nearly all body muscles!
    (b) Often leads to excessive consumption of fatty foods
    (c) Strongly interferes with the absorption of UV light
    (d) Makes bones too thick and strong for their own good!

10. The original stem or parent cell for all formed elements of the blood:
    (a) Lymphoblast
    (b) Megarkaryoblast
    (c) Platelet
    (d) Hemocytoblast

Body-Level Grids for Chapter 5

Several key body facts were tagged with numbered icons in the page margins of this chapter. Write a short summary of each of these key facts into a numbered cell or box within the appropriate *Body Level-Grid* that appears below.

Anatomy and *Biological Order* **Fact Grids for Chapter 5:**

TISSUE
Level

1

Physiology and *Biological Order* **Fact Grids for Chapter 5:**

CELL
Level

1

2

3

TISSUE
Level

1

ORGAN
Level

1

2

3

Physiology and *Biological Disorder* **Fact Grids for Chapter 5:**

CELL
Level

1

TISSUE
Level

1

ORGAN
Level

1

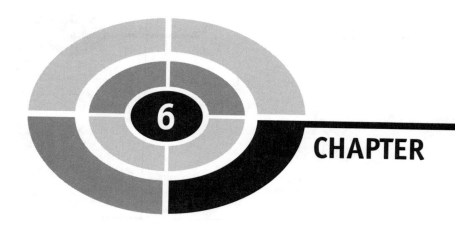

6

"Gentlemen, *Fire Up Your Engines!*": The Physiology of Neurons and Muscle Fibers

Chapter 5 covered the systemic physiology of the skin and skeleton. But it did not include the physiology of body movement. As soon as we get into body movement, both the *neurons* (nerve cells) and *muscle fibers* (thin, fiber-shaped muscle cells) become involved.

Lever Systems: Muscles, Bones, and Joints

A *bone–muscle lever* (**LEE**-ver) *system* is a combination of a bone, skeletal muscle, and movable joint that together carry out some particular body movement. Consider, for example, the *flexion* (**FLEK**-shun) or "bending" motion of the forearm (Figure 6.1).

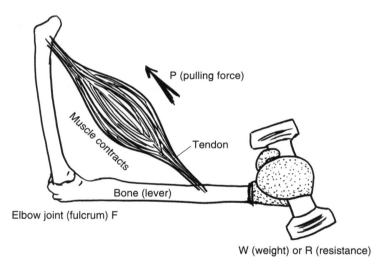

P (pulling force)

Muscle contracts

Tendon

Bone (lever)

Elbow joint (fulcrum) F

W (weight) or R (resistance)

Fig. 6.1   Flexion at the elbow: model of a bone–muscle lever system.

A lever is a rigid bar that is acted upon by some force. In a bone–muscle lever system, the bone serves as a passive lever or rigid bar that is acted upon by a *pulling force* (P). This pulling force is created by the contraction (short-ening) of a particular *skeletal muscle*. A skeletal muscle is a fleshy, reddish-colored organ that is attached to the bones of the skeleton.

For flexion of the forearm, the *biceps* (**BUY**-seps) *brachii* (**BRAY**-kee-**eye**) is the main skeletal muscle or prime mover. It attaches by means of *tendons* to the bones of the forearm. When the biceps brachii muscle contracts, it creates a pulling force (P) upon its lower tendons, which causes the bones of the lower arm to *flex* or "bend" at the elbow joint.

A *fulcrum* (**FULL**-crumb), abbreviated as F, is a place of "support" upon which a lever turns or balances when it moves. Within a bone–muscle lever system, a movable *joint* generally serves as the fulcrum. (A joint is simply a place of union or meeting between two bones.)

Finally, there is always some amount of *weight* (W) or *resistance* (R) that needs to be lifted during the body movement. In the present case, suppose the hand was holding a 5-pound dumbbell.

In summary, as the biceps brachii muscle contracts, it pulls upon its tendons. The tendons pull upon the bones of the lower arm, and the weight or resistance in the hand is lifted as flexion occurs at the elbow joint.

Motor Neurons and Muscle Fibers: The Neuromuscular Connection

Organ 1

Bone-muscle lever systems, to be sure, cause portions of the human body to move around. To review, the bones are the *passive* partners in body movement, because they merely serve as sticks or rigid bars that are pulled upon. The skeletal muscles, in contrast, are the *active* partners in body movement. The reason is that they are the organs that actively contract or shorten, providing the pulling force (P) upon their tendons, and finally, upon the bones.

IMPORTANCE OF THE MUSCLE FIBER (MUSCLE CELL)

The detailed gross anatomy of the skeletal muscle organ and its parts is talked about within *ANATOMY DEMYSTIFIED*. Here in our World of Physiology, however, our primary focus will be upon the functions of the individual skeletal muscle fiber. "Why is this, Professor Joe?", the physiologically inclined reader may ask of the Good Teacher. "It is due to the fact that the individual skeletal muscle fiber or cell, as well as groups of related muscle fibers, are the fundamental units of muscle contraction," is the reply.

"But do these muscle fibers, or groups of muscle fibers, normally contract on their *own*?", the astute reader quickly follows-up. "I see that you have actually been doing some *thinking* while you have been reading!," Joe responds. "The skeletal muscle fiber has to be *stimulated* – 'fired up' or 'turned on' – before it will contract."

ANATOMY AND PHYSIOLOGY OF THE MOTOR NEURON

"What part of the body stimulates, fires up, or turns on the muscle fiber, Professor?" Adjusting his glasses to think, Joe points to Figure 6.2. "The muscle fiber is turned on or excited by a particular *motor neuron*."

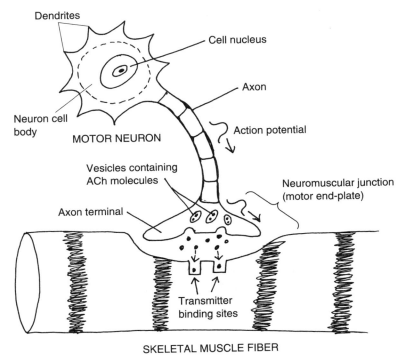

Dendrites

Cell nucleus

Axon

Neuron cell body

MOTOR NEURON

Action potential

Vesicles containing ACh molecules

Neuromuscular junction (motor end-plate)

Axon terminal

Transmitter binding sites

SKELETAL MUSCLE FIBER

Fig. 6.2   The neuromuscular junction and muscle fiber excitation.

Cell 1

A motor neuron, as its name suggests, stimulates one or more muscle fibers to contract and "move". Hence, we call it a "motor" (like the motor in your car) nerve cell. You might therefore think of a motor neuron as a type of self-exciting spark plug, which fires up the muscle fiber engine.

What are some of the parts of this natural spark plug? The microanatomy of the nerve cell centers upon its *cell body* or *neuron soma* (**SOH**-mah). The cell body (soma) is the major central portion of a neuron, which contains its nucleus and most other organelles. Arranged around the neuron soma or cell body are a number of *dendrites* (**DEN**-dryts). Like the branches of a "tree," the dendrites are slender branches of cytoplasm that carry information or excitation *toward* the cell body of the neuron.

An *axon* (**AX**-ahn) is a single branch of cytoplasm that resembles a long slender "axle" (like the axle found between the two wheels of a car). Unlike a car axle, though, the neuron axon doesn't actually move. Instead, it is a slender, fiber-shaped branch of cytoplasm that carries excitation *away from* the cell body of the neuron. (Since both the dendrites and the axon are slender and *fiber*-like, they are often collectively called *nerve fibers*.)

NERVE FIBERS	=	Axons	+	Dendrites
(thin, *fiber*-like branches of a neuron's cytoplasm)		(carry information *away from* neuron cell body)		(carry information *toward* neuron cell body)

The axon carries an *action potential* (also called *nerve impulse*) away from the cell body of the neuron. The action potential (nerve impulse) can be pictured as a wave – a traveling wave of *electrochemical* (e-**LEK**-troh-**chem**-ih-kal) or *ionic* (eye-**AH**-nik) excitation. It is mainly created as a large amount of sodium (Na^+) ions suddenly diffuse into the neuron through the channels in its plasma membrane (Chapter 4). Therefore, an action potential or nerve impulse represents ionic excitation, which "pertains to" (-*ic*) a lot of "ions" – the ions primarily being Na^+ cations. The term electrochemical literally "refers to" (-*al*) "chemicals" (such as sodium) that have a net "electrical" charge, such as the sodium ion, whose net (overall) charge is +1. Ions, such as the well-known sodium (Na^+) cation and the chloride (Cl^-) anion, can also be classified as electrochemical particles. (Review the discussion of ions in Chapter 3, if desired.)

Cell 2

ACTION POTENTIAL (NERVE IMPULSE)	=	A Traveling Wave of Electrochemical (Ionic) Excitation, Mainly Due to Massive Diffusion of Na^+ Ions into the Neuron

Eventually, the action potential travels down the axon to its branching end-tips, the *axon terminals*. Within these axon terminals are many tiny *vesicles* (**VES**-ih-kls). Each vesicle is literally a "tiny bladder" that consists of a membrane surrounding thousands of *neurotransmitter* (**NUR**-oh-**trans**-mit-er) *molecules*. By the term neurotransmitter is meant "sender across" (*transmitter*) of a "nerve" (*neuro-*) message. Each time an action potential reaches the axon terminals, the membranes surrounding some of the vesicles rupture.

Cell 1

Hundreds of neurotransmitter molecules are then released into the *neuromuscular* (**nur**-oh-**MUS**-kyoo-lar) *junction*. An alternative name for the neuromuscular junction is the *motor end-plate*. The neuromuscular junction (motor end-plate) is the flat, plate-like area where the axon terminals of a motor neuron almost (but not quite) touch the cell membrane of a nearby muscle fiber.

THE PROCESS OF MUSCLE FIBER EXCITATION

Once released, the neurotransmitter molecules diffuse across the narrow, saltwater-filled gap of the neuromuscular junction (motor end-plate).

Remember that we have defined a neurotransmitter as a "sender across of a nerve" message. In this case, the released molecules are being "sent across" the neuromuscular junction by diffusion. The molecules then attach to *transmitter binding sites* on the muscle fiber.

The type of message being "sent across" the neuromuscular junction is an *excitatory* (ek-**SIGH**-tah-tor-ee) one, because it is "characterized by" (*-ory*) "excitation." Specifically, the motor neuron was excited and fired a *neuron action potential*. As a result, eventually the vesicles in the axon terminals release *excitatory neurotransmitter molecules*. As their name reveals, excitatory neurotransmitter molecules stimulate or excite the cell to which they bind or attach.

Within the neuromuscular junction, the main excitatory neurotransmitter is called *acetylcholine* (uh-**see**-tul-**KOH**-leen), abbreviated as *ACh*. "How do we know that the muscle fiber or cell is excited, Professor Joe?" The answer to this question is obvious: the muscle fiber, when it is excited, has a *muscle fiber action potential* of its own, which is created in the region of the neuromuscular junction. In summary of action potentials, here is a general rule of thumb:

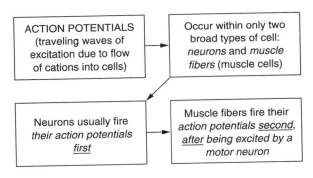

Microanatomy of a Skeletal Muscle Fiber

In order to fully understand how a particular muscle fiber contracts after it is stimulated or excited at the neuromuscular junction, we need to do some further work on the microanatomy of the muscle fiber, itself. Figure 6.3 gets us started. A thin plasma membrane, called the *sarcolemma* (**sar**-koh-**LEM**-uh) or "flesh" (*sarc*) "husk" (*lemm*), surrounds the outside of the muscle fiber. Skeletal muscle fibers bear dark cross-stripes, called *striations* (**stry-AY**-shuns).

Why is the skeletal muscle fiber cross-striped? To show why, we look at the interior of the fiber. Here are a number of *myofibrils* (**my**-uh-**FEYE**-brils).

154

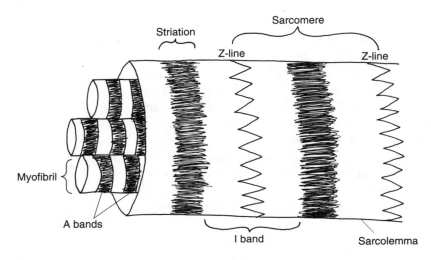

A MUSCLE FIBER (CELL)

Fig. 6.3 A look within a muscle fiber (cell).

The word myofibril literally means "little fiber" (*fibril*) of a "muscle" (*my*). In reality, however, each myofibril is just a long, thin, fiber-shaped organelle within a muscle fiber. The myofibrils have a dark-and-light *banding pattern*. The dark bands are called the *A bands*, while the light bands are called the *I bands*. The striations (cross-stripes) of the entire muscle fiber, therefore, are created by the stacking of the dark A bands of the myofibrils, one upon another.

Within the middle of each light I band is a dark, zig-zagging *Z-line*. These dark, jagged lines mark off a series of *sarcomeres* (**SAR**-koh-**meers**). A sarcomere is a short "segment" (-*mere*) of "flesh" (*sarc*), that is, a region of myofibril between two Z-lines. Hence, each myofibril organelle within a muscle fiber basically consists of a series of sarcomeres, attached end to end.

A LITTLE WIENIE WITH HOLES: TRANSVERSE TUBULES AND THE SARCOPLASMIC RETICULUM

Now that we have talked about some of the muscle fiber's microanatomy, we are ready to dig deeper into its "flesh," which in Greek is denoted by the word root *sarc*. Speaking of flesh, it would be useful to visualize each skeletal muscle fiber as a little 3-D wienie or hot dog, with black cross-stripes (striations) on it, from being cooked on an outdoor barbecue grill! We need to think three-dimensionally, so that we can grasp the fact that the muscle fiber is not flat, but rather is a solid cylinder of flesh that *contains* various organelles and complex groups of molecules.

Take a wienie and stick toothpicks all the way "across" (*trans-*) and through it. Pull the toothpicks out. What are you left with? You have a series of *transverse tubules*, also called *T-tubules*. The T (transverse) tubules are "tiny tubes" (*tubules*) that go all the way through and across the muscle fiber, from one side of it to the other (Figure 6.4).

Flanking either side of the T-tubule are the *lateral sacs* of the *sarcoplasmic* (sar-koh-**PLAZ**-mik) *reticulum* (reh-**TIK**-you-lum). A reticulum, in general, is "a little network" (*reticul*) that is "present" (*-um*) within a cell. In muscle fibers, which, as we have pointed out, consist of "flesh" (*sarc*) or solid "matter" (*plasm*), the reticulum is given a special name. The sarcoplasmic reticulum is a little network of hollow sacs and tubules that branches extensively

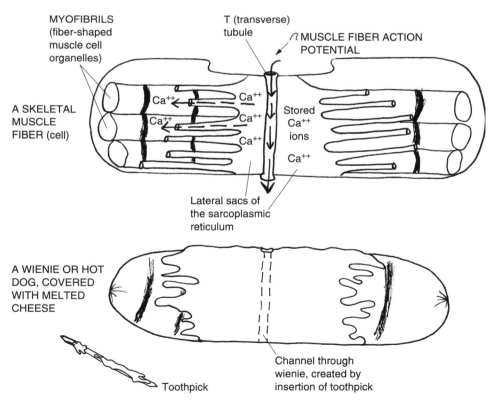

Fig. 6.4 A little wienie with holes and cheese: the lateral sacs, T-tubules, and sarcoplasmic reticulum.

within the cytoplasm of the muscle cell. [**Study suggestion:** It may seem a bit weird, but picture the little hot dog or wienie as having Swiss cheese melted over it. The cheese is stringy and creates a gooey "network" of sorts. What particular organelle is the melted, stringy Swiss cheese modeling?]

The lateral sacs serve as storage depots for calcium (Ca^{++}) ions within the muscle fiber. When a muscle fiber action potential is created in the region of the neuromuscular junction, it flows down into the T-tubule, which carries it all the way through the muscle fiber. As the action potential travels, it temporarily makes the lateral sacs of the sarcoplasmic reticulum much more leaky to calcium ions. As a result, thousands of Ca^{++} ions leak out of the lateral sacs, and into the region of the myofibrils.

Muscle Fiber Contraction: The Sliding Filament Theory

"Where are all these calcium ions going to, within each myofibril?," the inquisitive brain yearns to know. As we mentioned earlier, an adequate concentration of calcium ions within the bloodstream (and stored inside the lateral sacs of the sarcoplasmic reticulum) is absolutely essential for the excitation and contraction of muscle fibers! We can understand why this is so if we dig really deep into the myofibril organelles packed into each muscle fiber. Here is a summary of a previous statement:

Organelle 1

A MYOFIBRIL (fiber-shaped muscle = A series of sarcomeres, attached
          cell organelle)   end to end

Since each myofibril is essentially a bunch of sarcomeres connected together, it is only logical that we should look to the individual sarcomere to discover *how* the entire myofibril functions.

MYOFILAMENTS WITHIN EACH SARCOMERE

A very powerful electron microscope can be used to magnify the sarcomere, tens of thousands of times! What do we see when we do this? Study Figure 6.5. Here we find ourselves back down at the Chemical Level of Body Organization. The high end of the Chemical Level, just below the level of the cell organelles, consists of *integrated groups of molecules.*

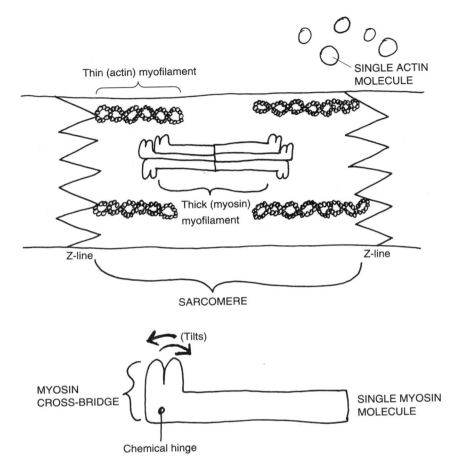

Fig. 6.5   A sarcomere and its myofilaments.

In a skeletal muscle fiber, these integrated groups of molecules are mainly functional groups of proteins. Of great importance are the *myofilaments* (**my-oh-FIL**-ah-ments) or "muscle" (*myo-*) "threads" (*filaments*). The myofilaments are actually thread-like collections of protein molecules. There are two types of myofilaments: *thin actin* (**AK**-tin) *myofilaments* and *thick myosin* (**MY**-oh-sin) *myofilaments*.

Molecule 1

The thin actin myofilaments are very long, slender threads consisting of two twisted strands of globe-shaped *actin proteins*. [**Study suggestion:** Find a beaded pearl necklace and lay it upon a surface. Place both strands of the necklace side by side, then twist them around each other. The resulting double helix (two twisted strands) provides a rough model for a thin actin myofilament.] The thin actin myofilaments are attached to the Z-line, at either end of the sarcomere.

Located right in the middle of each sarcomere is a series of thick myosin myofilaments. These are stacked vertically above and below one another, with narrow gaps between them. Each myosin myofilament consists of dozens of individual myosin protein molecules. Each myosin protein somewhat resembles a golf club with two heads. The double-heads are tiltable (using energy from split ATP). It is as if the myosin double-head were poised upon a chemical hinge. The tiltable double-head of each myosin molecule is technically called a *myosin cross-bridge*. [**Study suggestion:** Visualize two golfers, each carrying their own bag of golf clubs. Each club has a double-head at one end, which is attached by a tiltable hinge. The two golfers stand back to back, in the center of a sarcomere, and then each gives his golf bag a heave. If the golfers keep hold of their bags, their clubs will come flying out in both directions, some with their double-heads pointing upward, and some with their double-heads pointing down. The resulting highly orderly arrangement provides a rough model for the thick myosin myofilament.]

The myosin cross-bridge is the chief contact point between the thin actin and thick myosin myofilaments. It is also crucially important because of its close functional relationship with the high-energy ATP molecule. You may recall (Chapter 4) that the ATP molecule is split by a special kind of enzyme, called ATPase. In the case of muscle, the enzyme is known as *myosin ATPase*.

CALCIUM IONS ALLOW INTERACTION BETWEEN MYOFILAMENTS

Let us get back to the calcium ions. After the muscle action potential travels through the T-tubule, memory may prod you (back in Figure 6.4), Ca^{++} ions diffuse out of the lateral sacs and into the region of the myofibrils. The calcium ions enter the sarcomeres of the myofibril. Once there, they combine with a group of proteins. One group of three globe-shaped proteins is called the *troponin* (**TROH**-poh-nin) *complex*. The other long, strand-like protein is named *tropomyosin* (**troh**-poh-**MY**-oh-sin).

The structure of troponin-tropomyosin

Troponin complex and tropomyosin are bonded together to create an *inhibitory protein group*. By inhibitory, we mean that troponin–tropomyosin *inhibits* or prevents a chemical connection between the thin actin and thick myosin myofilaments. How do they prevent this connection? Well, the tro-ponin–tropomyosin group is like a dark, leather, double-twisted strand (the long, thin tropomyosin molecule) with a few dark beads (the rounded tro-ponin molecules) attached along the sides of it. This dark leather strand–dark bead combination looks like a pair of snakes, each with three buggy eyes, lying within the grooves of the double-helix molecule of the actin protein pearl necklace! (Consult Figure 6.6, A.) We can summarize the total structure of the thin myofilament as:

A

THIN MYOFILAMENT	= A Big Double Helix of *Actin* Molecules (like a white, twisted pearl necklace)	+ A Smaller Double Helix of Dark *Tropomyosin*, with Dark Beads of *Troponin* attached, lying like two bug-eyed snakes inside the long grooves of the Actin Double Helix

Molecule 2

Tropomyosin covers up the actin; troponin combines with calcium

We will add further to our snake analogy. We can say that tropomyosin, like a long, dark, sneaky snake, covers up the *binding sites for myosin* that are present on the pearl-like beads of actin! (Study Figure 6.6, B.) "So, I guess you're saying that tropomyosin is some kind of sneaky trouble-maker, Professor, at least as far as keeping actin and myosin apart?" Well, that's pretty close!

The muscle fiber stays in a *state of relaxation*, with its contraction being *inhibited*, while tropomyosin is covering up the actin molecules' binding sites for myosin.

In order to finish the process of muscle excitation, this inhibition of muscle contraction by the tropomyosin must be removed or relieved. Now, *this* is where the calcium ions finally come in! After the muscle fiber action potential releases stored Ca^{++} ions, they enter the sarcomeres of the myofibrils and bond to the buggy-eyed *troponin protein complex*. (View Figure 6.6, C.) With those calcium ions stuck to its buggy troponin eyes, the sneaky snake (long skinny tropomyosin molecule) changes its shape, breaking away from the actin molecules and falling deeper down into the groove.

At last! The sneaky snake (tropomyosin) no longer covers the binding sites for myosin! Hence, tiny projections on each double-headed myosin cross-bridge now fit into the binding sites on the actin pearl beads. (Examine

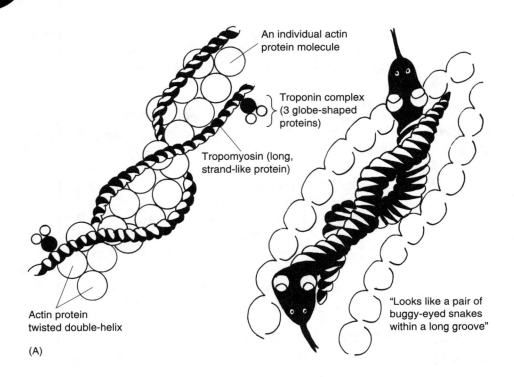

An individual actin
protein molecule

Troponin complex
(3 globe-shaped
proteins)

Tropomyosin (long,
strand-like protein)

Actin protein
twisted double-helix

(A)

"Looks like a pair of
buggy-eyed snakes
within a long groove"

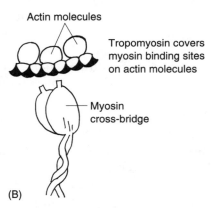

Actin molecules

Tropomyosin covers
myosin binding sites
on actin molecules

Myosin
cross-bridge

(B)

Fig. 6.6 Troponin–tropomyosin: buggy-eyed snakes in a groove! (A) Troponin–tropomyosin and the actin protein double helix. (B) Close-up showing troponin–tropomyosin covering the binding sites for myosin on surfaces of actin molecules. (C) Free Ca^{++} ions attach to the troponin–tropomyosin, breaking tropomyosin away from the actin molecules. (D) Myosin cross-bridge fits into exposed binding sites on actin molecules, completing actin–myosin coupling.

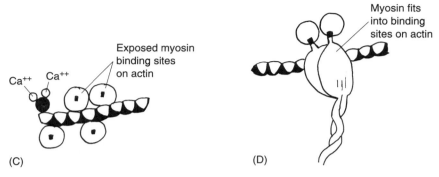

Fig. 6.6 (continued)

Figure 6.6, D.) Actin and myosin have met and bonded. Muscle fiber excitation is now complete, and we're ready to contract!

READY, SET, *SLIDE!*

According to the *sliding filament theory of muscle contraction*, muscles shorten due to the inward sliding of the thin actin myofilaments over the tilted cross-bridges of the thick myosin myofilaments. "How does this sliding process get started, Joe?" The answer involves the action of the enzyme, myosin ATPase.

Molecule 1

After actin has combined with myosin, the myosin ATPase enzyme becomes activated. It splits ATP molecules in the region of the myosin cross-bridges. The resulting free energy tilts the myosin cross-bridges inward. As the cross-bridges tilt, the overhanging thin actin myofilaments slide over their tips. (View Figure 6.7.) The sliding occurs at both ends of the sarcomere. Hence, each sarcomere shortens. Since each myofibril organelle consists of a series of sarcomeres hooked end to end, the whole myofibril shortens. And as all of their myofibril organelles shorten, the entire muscle fiber (cell) also shortens.

GETTING RID OF CALCIUM: THE PROCESS OF RELAXATION

It requires splitting of ATP to provide the energy for tilting of the myosin cross-bridges during muscle contraction. Strangely enough, ATP is also required for *relaxation*! The reason: we must get rid of the Ca^{++} ions, if

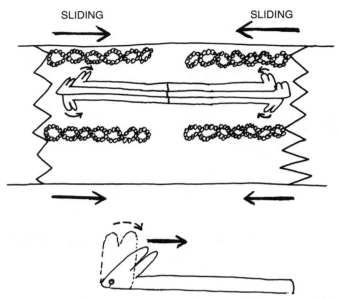

Fig. 6.7. Muscle fiber contraction: an inward sliding of the thin myofilaments.

Molecule 1

relaxation is to occur. Otherwise, as in *rigor* (**RIH**-gur) *mortis* (**MOR**-tis) or "death stiffness," the actin molecules and the myosin cross-bridges remain locked in a state of contraction.

There is an *active transport calcium pump system* (Chapter 5) located within the muscle fiber. It is constantly splitting ATP and obtaining free energy that actively transports (pumps) Ca^{++} ions away from their connection to troponin, and back into the lateral sacs of the sarcoplasmic reticulum. After Ca^{++} has been pumped away, the sneaky snake, tropomyosin, flips back into resting position and once again covers the myosin binding sites on the actin molecules.

The cross-bridges tilt back again into their vertical positions. This causes the thin actin myofilaments to slide back outward. The sarcomere re-lengthens, and the muscle fiber relaxes.

E–C–R: THE MUSCLE ACTIVITY CYCLE

This same highly orderly sequence of muscle excitation (E), contraction (C), and relaxation (R) occurs again and again throughout our lives. Small wonder, then, that it is called *E–C–R: The Muscle Activity Cycle*.

Motor Units and the Concept of Threshold

"No man is an island unto himself," so the old saying goes. The same general statement could be reworded and applied to what we have learned about the physiology of the skeletal muscle fiber. "No skeletal muscle fiber is an island unto itself. Rather, it functions as part of a particular *motor unit*."

A motor unit consists of a certain motor neuron, and all of the skeletal muscle fibers that it supplies. Way back in Figure 6.2, we illustrated the essential components of the neuromuscular junction (motor end plate). Its layout was correct, but it was just oversimplified. In reality, the axon of a particular motor neuron branches extensively.

Thus, a single motor neuron *innervates* (**IN**-er-**vayts**) or sends "nerve" (*nerv*) fibers "into" (*in-*) more than just one muscle fiber. Figure 6.8 shows two *lower motor neurons* located within the *horns of gray matter* in the *spinal cord*. On average, a single lower motor neuron innervates 150 skeletal muscle fibers! Hence, we are not just dealing with a single neuromuscular junction, but with over one hundred of them!

Note in Figure 6.8 that lower motor neuron A supplies or innervates skeletal muscle fibers 1 and 2. Its next-door neighbor, motor neuron B, innervates skeletal muscle fibers 3, 4, and 5. Realize that these examples

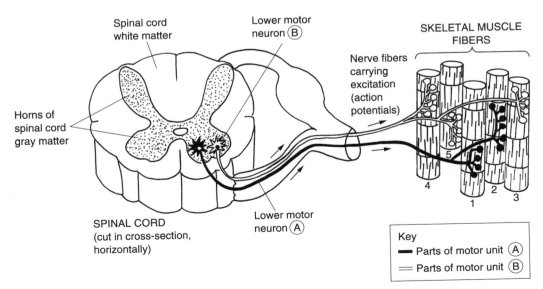

Fig. 6.8 Two motor units supplying a skeletal muscle organ.

are highly simplified, in that the real-life situation would involve hundreds of muscle fibers within a particular muscle organ.

Returning to the idea of a motor unit, we can identify two different motor units in Figure 6.8. *Motor unit A* consists of motor neuron A plus skeletal muscle fibers 1 and 2. However, *motor unit B* involves motor neuron B plus skeletal muscle fibers 3, 4, and 5.

THRESHOLD, TWITCH, AND THE "ALL-OR-NONE LAW"

"Why do we bother with identifying particular motor units in a certain skeletal muscle organ, Professor Joe?" The answer is that it helps us understand the real-world contraction of skeletal muscle organs. Normally, of course, we would expect a particular skeletal muscle (such as the biceps brachii in the upper arm) to respond and contract, if it was excited with a stimulus that was sufficiently strong.

But an important question to ask ourselves here is, "Every time that the biceps brachii muscle contracts, does the lower arm always flex with the same amount of force?" The answer, of course, is "No." The contractions of a particular skeletal muscle organ are *graded* in their strength or intensity. What we mean by "graded" is that the contractions can become progressively stronger, or progressively weaker, as if they were going up or down a *graded* hill, or a series of higher and lower steps.

It takes very little mental effort and strength of biceps brachii contraction, for instance, to flex the arm and lift a piece of paper. It takes more mental effort, and a greater force of biceps contraction, to raise a heavy book. And it takes supreme mental effort, and a near-maximal force of biceps contraction, to lift a dumbbell weighing 100 pounds (with one hand)! These whole-muscle contractions, then, are graded in their strength or force.

These observations lead us to state what we can call *The Motor Unit Rule:*

All muscle contractions are due to the contractions of a certain number of motor units within the muscle. The greater the number of motor units contracting, and the more skeletal muscle fibers these motor units contain, the greater will be the resulting force of muscle contraction.

The concept of threshold

"Okay, Professor Joe, but how do we get even *one* motor unit to contract?" You have to stimulate it strongly enough to reach its *threshold*. A threshold, in general, is a place where you "tread" or "tramp," a point where you cross or enter a new beginning point (as in crossing the threshold of a doorway and entering a house). For our purposes, the threshold of a skeletal muscle is the lowest strength or voltage of stimulation required to get contraction of at least one of its motor units. Let us say that you artificially stimulate a skeletal muscle by inserting electrodes into it. You turn on an electric current, and zap the muscle with a shock of 1 *volt* (V). If nothing happens, then this 1-volt shock is a *subthreshold stimulus*. It is too weak and falls "below" (*sub-*) the intensity of stimulation required to get at least one motor unit to contract.

What to do? Obviously, we keep raising the strength or voltage of electrical stimulation, and keep delivering more powerful shocks. Suppose after delivering a shock of 2 volts, we get a noticeable contraction response of the muscle. We can thus call this shock a *threshold stimulus*.

Muscle twitches

We can see the precise behavior of the muscle for ourselves by employing an instrument called a *myograph* (**MY**-oh-graf) – "an instrument used to record" (*-graph*) the movements of a "muscle" (*my*). Figure 6.9 displays the chart tracings called a *myogram* (**MY**-oh-gram) or "graphical record" (*-gram*) of "muscle" (*my*) activity.

At a threshold stimulus of, say, 2 volts, we get a very small *muscle twitch*. A muscle twitch is a quick, jerky contraction of one or more motor units whose stimulus threshold has been reached. Remembering the E–C–R sequence of the Muscle Activity Cycle, the E (excitation) is the threshold stimulus of 2 volts. A small number of motor units contract (C), raising the tracing on the myogram up from its horizontal baseline.

The muscle twitch is not very effective in human body movements, however, because almost immediately after the contraction phase, relaxation (R) occurs. The myogram record of the twitch then falls back to its horizontal baseline starting level. In intact humans, the single twitch often occurs as a *muscle tic*. A muscle tic is a quick, jerky spasm of a muscle (as on your cheek or eyelid). It is often due to some local irritation of the nerve endings or muscle fibers in a body region.

The All-or-None Law

Glance back at Figure 6.8. Pretend that on the myogram recording of the muscle twitch at 2 volts (Figure 6.9), it was motor unit A and its two skeletal muscle fibers (1 and 2) that responded. Note that there is a law, called *the All-*

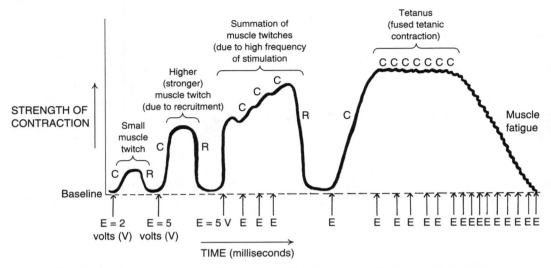

Fig. 6.9 A myogram record of a muscle artificially stimulated with electrical shocks.

Tissue 1

or-None Law, that describes the behavior of a motor unit. According to the All-or-None Law, either *all* of the muscle fibers in a given motor unit contract together at the same time, or *none* of them contract. So, for example, we cannot have just skeletal muscle fiber 1 contract, while 2 just sits there like a dead horse! Whenever motor unit A is zapped by a threshold stimulus, then all of its muscle fibers (both 1 and 2) contract. In summary, all of the skeletal muscle fibers in a given motor unit contract together, at the same time. And as their name indicates, they "move" (motor) together as a single "unit."

RECRUITMENT, SUMMATION, AND TETANUS

The first part of the myogram shown in Figure 6.9 represented a minimum-sized muscle twitch. The twitch wasn't very strong, because it was the contraction of just one or a few of the motor units in a muscle – the ones with the lowest threshold (or greatest sensitivity) to stimulation. In our example back in Figure 6.8, we said that motor unit A was the more sensitive, so that its fibers 1 and 2 would contract when given a weak shock of just 2 volts.

"What happens if we greatly raise the strength of muscle stimulation, Professor Joe?" If we increase the voltage of our electrical stimulator to 5 volts, then we get a significantly higher muscle twitch (Figure 6.9). This higher twitch reflects the *recruitment* of motor units. Recruitment is the bringing into action of more and more motor units within a muscle, as the strength of muscle stimulation is increased. The main reason for recruitment is that you are reaching the threshold of a greater and greater percentage of motor units in a muscle, as you increase the voltage of stimulation. Suppose, for instance, that motor unit B (back in Figure 6.8) has a higher threshold than motor unit A. If its threshold is, say, 4 volts, then it, too, will be recruited and start contracting (along with motor unit A) when a shock of 5 V is delivered. If the voltage of single shocks keeps getting higher and higher, eventually there will be a 100% recruitment of all of the motor units in a muscle, and a *maximum twitch* will result. [**Study suggestion:** When motor unit B is recruited, why would you expect the resulting twitch to be a lot stronger than the one produced by a shock of only 2 volts? Glance back at Figure 6.8, if desired, to help you reason out the answer.]

Summation of muscle twitches

Now that we understand something about muscle twitches, and recruitment of motor units at higher levels of single shocks, it is time to look at the effect of increasing the *frequency* of muscle stimulation. Let us go back to a single shock of 5 volts. With one such shock, both motor units A and B (and skeletal muscle fibers 1–5) will be recruited, producing a fairly strong muscle twitch. But as we have already pointed out, individual muscle twitches have a much too rapid onset of relaxation to allow them to be of much practical use for body movements.

The key strategy used by the body to move its parts, then, is to create a *summation of muscle twitches*. As pictured in the myogram of Figure 6.9, summation is the *adding together* of muscle twitches, due to a *high frequency* of muscle stimulation. Let us make a homely analogy. Mom yells at Baby Heinie to get off the couch and clean up the living room. If Mom yells just once (loudly enough to reach the kid's *sensory threshold*), then Baby Heinie will just jump off the couch once, pick up one of his toys, and then immediately collapse and relax back on the couch! The room is still a mess with many scattered toys! A much better strategy (well known by all care-worn Moms and Dads) is to yell at the kid once, and (before he has had a chance to relax and flop back down) yell at him again and again!

Tissue 2

This creates a summation effect, because Baby Heinie (just like the involved motor units) is stimulated to contract and move again, *before* he has had a chance to completely relax and lengthen his sarcomeres! The result is a fairly continuous body movement, one that can be used to accomplish practical tasks like handing people a piece of paper, or picking toys up off the floor. Therefore, most of our body movements represent the summation of muscle twitches.

Tetanus (tetanic contractions)

"What happens if the muscle is stimulated at an extremely high frequency, say at a rate of 80 to 100 shocks every second?" The myogram back in Figure 6.9 displays the response as *tetanus* (**TET**-ah-nus) or, using more appropriate terminology, as a *fused tetanic* (**teh-TAN**-ik) *contraction*. Both the words tetanus and tetanic derive from the Greek word root *tetan*, which means "extreme tension" or "convulsive tension." In normal muscle physiology, tetanus (tetanic contraction) is one long, sustained muscle contraction, with no relaxation whatsoever, due to an extremely high frequency of muscle stimulation.

Figure 6.9 shows this myogram as a *fused tetanus*, since the individual contractions of the motor units have all been *fused* together into a single, "extreme tension," flat, plateau-like response. Tetanus (tetanic contractions) occur whenever we consciously keep stimulating our muscles, not allowing ourselves to relax at all!

Muscle fatigue

Organ 1

"Can tetanus or tetanic contractions last forever, Professor? What if we just keep stimulating the muscle – on and on and on?" Well, we know that no physiological activity of the human body can last forever, don't we? The same is true for tetanus. Eventually a state of *muscle fatigue* sets in. Fatigue is literally "a tiring," due to exhaustion of available energy and a build-up of metabolic waste products.

Muscle Endurance Versus Fatigue: Biochemical Fiber Types

Well, did you figure out that muscle fatigue is mainly due to a using up of all available ATP energy within the muscle fibers? – Good for you! And we can logically conclude that avoiding muscle fatigue, whenever possible, has definite survival advantages for the human body! Pretend, for example, that you are a primitive hunter-gatherer, living long before civilization. You need good endurance to keep tracking the deer you speared, or to keep running away from a snarling bear! At other times you just need to make a quick, mad dash for it, and not worry about muscle fatigue.

In short, we can see a physiological pattern of muscle fiber activity in operation, here. This physiological pattern reinforces the ideas of Walter Cannon (Chapter 3) about the "fight-or-flight response" and his book titled *The Wisdom of the Body*. Let us now state a natural rule that we can imaginatively call *Muscle Fiber Wisdom*:

Muscle Fiber Wisdom: All of the skeletal muscle fibers within a particular motor unit are of a certain **biochemical** *type.* Thus, our skeletal muscle fibers have their own biochemical "wisdom." This allows particular types of muscle fibers to take the lead in responding to particular exercise-demanding conditions.

MUSCLE FIBER ENERGY STORAGE

There are three forms of potential energy storage within all skeletal muscle fibers:

1. Muscle fiber ATP. There is only a small quantity of ATP molecules stored within the muscle fiber. It provides a short-term energy supply for both contraction and relaxation.

2. Creatine phosphate. A longer-term form of energy storage is offered by molecules of *creatine* (**KREE**-ah-tin) *phosphate*, abbreviated as *CP*. Creatine is literally "a substance" (*-ine*) within "flesh" (*creat*), that is, muscle. Creatine phosphate, like adenosine triphosphate (ATP), contains high-energy chemical bonds with phosphate. These high-energy bonds can readily be broken, resulting in free (kinetic) energy. This energy puts phosphate (P) onto ADP, thereby generating more muscle ATP.

There is about 3 to 6 times as much CP (creatine phosphate) stored within skeletal muscle fibers, compared to ATP! Thus, creatine phosphate (CP) serves as a longer-term energy storage that helps quickly produce more ATP, thereby postponing muscle fatigue. (CP cannot be used by the muscle for energy; it just helps make ATP.)

3. Muscle fiber glycogen. Glycogen ("sweetness or glucose producer"), you may remember (Chapter 3), is basically a large carbohydrate molecule consisting of many bonded glucose molecules. It thus represents a form of stored glucose (and its potential energy) that can be tapped within the muscle fiber.

RED (HO) FIBERS VERSUS WHITE (LO) FIBERS

Human skeletal muscle organs are mixtures of two basic colors or shades of muscle fibers. There are *red or dark fibers* mixed with *white or light fibers*. The red (dark) muscle fibers owe their color to a high concentration of *myoglobin*

(**MY**-oh-**gloh**-bin). Myoglobin is a "globe"-shaped (*glob*) "muscle" (*myo*) "protein" (-*in*).

Molecule 2

Myglobin is reddish in color. Hence, red (dark) muscle fibers have a high myoglobin concentration, while white (light) muscle fibers have a low myoglobin concentration. Myoglobin mainly functions as a temporary store for oxygen (O_2) molecules within the muscle fiber. (Each myoglobin molecule temporarily attaches and stores one O_2 molecule.) It thereby helps the red or dark fiber contract aerobically, using oxygen. Red fibers are often classified as *HO* (high oxidative) or aerobic muscle fibers.

In addition, red (HO) fibers have a high density or crowding of mitochondria within their cytoplasm. This makes them quite resistant to fatigue, since the aerobic or oxidative catabolism of glucose (by the Krebs cycle and electron transport system) produces a whopping net of 38 ATPs/glucose. Recall (Chapter 4) that cellular respiration or aerobic breakdown of glucose occurs within the mitochondria.

White fibers, in contrast, contain very little myoglobin. Their mitochondrial density is low. They are classified as *LO* (*low oxidative*) or anaerobic muscle fibers. They chiefly break down glycogen and glucose by glycolysis in the cytoplasm, which requires no input of oxygen. Anaerobic (LO) fibers produce ATP quickly, because glycolysis requires very few biochemical steps. But they are pretty quick to fatigue, since they only get a net of 2 ATPs produced/glucose.

FAST TWITCH (FT) VERSUS SLOW TWITCH (ST) FIBERS

Another way of classifying muscle fibers biochemically is according to their *fiber twitch speed* (speed of twitching or contracting). Fiber twitch speed is basically determined by a muscle fiber's concentration of the enzyme, myosin ATPase. Recall that myosin ATPase is the enzyme that splits ATP near the myosin cross-bridges, thereby releasing free energy for tilting and sliding the thin actin myofilaments. In general, the higher a skeletal muscle fiber's concentration of myosin ATPase, the faster it will split ATP, and the faster the fiber will twitch.

Fast twitch or *FT muscle fibers* thus have a high concentration of myosin ATPase. They produce quick contractions, but they are more likely to fatigue, since they use up ATP at a fast rate.

Slow twitch or *ST muscle fibers*, in contrast, have a low concentration of myosin ATPase enzyme. Their contractions are fairly slow. But their payoff is better endurance, since they are breaking down ATP at a slow rate.

Cell 3

COMBINING CHARACTERISTICS: THE THREE SKELETAL MUSCLE FIBER TYPES

There are three biochemical skeletal muscle fiber types. Each of these types is a combination of two factors: oxidative or aerobic capacity (HO versus LO) and myosin ATPase concentration (FT versus ST).

Cell 4

Fiber type 1: HOST (High Oxidative, Slow Twitch)

These fibers are red and aerobic. Thus, they are very efficient producers of ATP. They twitch slowly, so they use up ATP at a slow rate. Hence, HOST fibers are generally the best for slow endurance activities, because they are the most resistant to fatigue. Picture HOST as the penny-pinching miser or Scrooge of the muscle fiber world – it has lots of "money" (ATP) but just won't "spend" it very fast! HOST fibers play an important role in your *postural* (**PAHS**-chur-al) *muscles*, such as those in your neck and lower back, that allow you to maintain an upright posture all day long without collapsing with fatigue!

Fiber type 2: HOFT (High Oxidative, Fast Twitch)

Like the HOST fibers, these are red, aerobic, and efficient ATP-producers. But since they have a high concentration of myosin ATPase enzyme, they split ATP at a rapid rate. These fibers are pretty good for endurance, yet still twitch rapidly. Picture HOFT fibers as the high-rollers or Yuppies of the muscle fiber world – they have lots of "money" (ATP) and they "spend" it rapidly (twitch fast). For example, these fibers would be important in the leg muscles of the winner of a 26-mile marathon race. The winner of the marathon needs good endurance to finish this long distance, but at the same time must avoid early fatigue.

Fiber type 3: LOFT (Low Oxidative, Fast Twitch)

Unlike HOST and HOFT, these fibers produce ATP anaerobically, without use of oxygen. They are thus the white muscle fibers, and the quickest fibers, because they produce 2 ATPs per glucose using just a few biochemical steps of glycolysis. And since they are fast-twitch (FT), they split their quickly produced ATPs at a rapid pace. Picture LOFT fibers as the energy spendthrifts of the muscle fiber world – they don't "earn money" (produce ATP)

very efficiently, but boy can they "spend it" (use up available ATP) very quickly! Hence, you would vote these skeletal muscle fibers as "most likely to *fatigue* (not *succeed*)"!

"Does this mean that LOFT fibers *aren't* very useful, Professor?" Not at all! The LOFT fibers get things done in a big hurry, when no one is worried about fatigue! In a sports context, think about the leg muscles in the winner of a 50-yard dash! The most genetically "gifted," blazing-fast runners have an unusually high percentage of LOFT fibers within their leg muscles!

[**Study suggestion:** Look back over the three biochemical fiber types and their characteristics. Try to match them up with the personalities of different people you know. Which one best describes *you*? Are you the energy miser or Scrooge (HOST), the Yuppie or high-roller (HOFT), or the bankrupt energy spendthrift (LOFT)?]

Quiz

Refer to the text in this chapter if necessary. A good score is at least 8 correct answers out of these 10 questions. The answers are listed in the back of this book.

1. The lever in any bone–muscle lever system is best represented by the:
   (a) Skeletal muscle
   (b) Bone
   (c) Joint
   (d) Pulling force

2. The nerve–muscle joining place is technically called the:
   (a) Knee joint
   (b) Axon terminal
   (c) Neuron soma
   (d) Neuromuscular junction (motor end-plate)

3. A traveling wave of electrochemical (ionic) excitation:
   (a) Sarcolemma
   (b) Dendrite
   (c) Action potential (nerve impulse)
   (d) Transverse tubule

4. A little network of hollow sacs and tubules within a skeletal muscle fiber:
   (a) Sarcomeres

    (b) Striations
    (c) Z-lines
    (d) Sarcoplasmic reticulum

5. Explains the phenomenon of muscle contraction:
    (a) Sliding filament theory
    (b) Hematopoiesis
    (c) Motor unit
    (d) Actin protein

6. Enzyme directly responsible for providing the energy for muscle contraction:
    (a) Lactic anhydrase
    (b) Myoglobin
    (c) Myosin ATPase
    (d) Troponin–tropomyosin

7. A dead skeletal muscle fiber can't relax, because:
    (a) Too much energy has pumped away all of the calcium
    (b) Its contractions are always isotonic
    (c) Actin and myosin remain locked together
    (d) No Na^+ ions are available

8. Every time skeletal muscle fiber X contracts, skeletal muscle fiber Y also contracts. From this information we can conclude that:
    (a) Both X and Y are part of the same motor unit
    (b) X is slow-twitch, while Y is fast-twitch
    (c) X is high-oxidative, while Y is low-oxidative
    (d) Neither fiber has an adequate energy source

9. A shock of 50 volts produces a twitch 10 millimeters high. A shock of 100 volts yields a twitch 18 millimeters high. Therefore:
    (a) Summation is occurring
    (b) Fatigue is setting in
    (c) Tetanus has been reached
    (d) Recruitment has been observed

10. An alternative, high-energy source stored within all muscle fibers:
    (a) Vitamin K
    (b) Ascorbic acid
    (c) Creatine phosphate
    (d) Myoglobin

Body-Level Grids for Chapter 6

Several key body facts were tagged with numbered icons in the page margins of this chapter. Write a short summary of each of these key facts into a numbered cell or box within the appropriate *Body-Level Grid* that appears below.

Anatomy and **Biological Order** Fact Grids for Chapter 6:

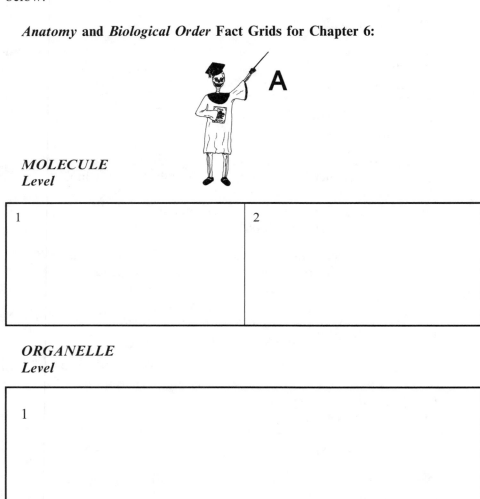

MOLECULE
Level

1	2

ORGANELLE
Level

1

CELL
Level

1

Physiology and *Biological Order* Fact Grids for Chapter 6:

CELL
Level

1	2
3	4

TISSUE
Level

1	2

ORGAN
Level

1

Physiology and *Biological Disorder* Fact Grids for Chapter 6:

ORGAN
Level

1

Function and *Biological Order* Fact Grids for Chapter 6:

MOLECULE
Level

1	2

Function and *Biological Disorder* Fact Grids for Chapter 6:

MOLECULE
Level

1

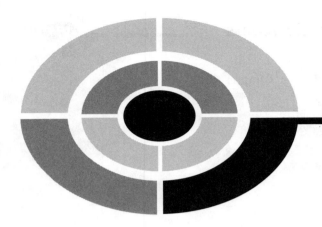

Test: Part 3

DO NOT REFER TO THE TEXT WHEN TAKING THIS TEST. A good score is at least 18 (out of 25 questions) correct. Answers are in the back of the book. It's best to have a friend check your score the first time, so you won't memorize the answers if you want to take the test again.

1. The skin or integument consists of:
   (a) Malleus, incus, and stapes
   (b) Epidermis, osteons, and subdermis
   (c) Squamae, epidermal strata, and osteocytes
   (d) Epidermis, dermis, and hypodermis
   (e) Subdermis, hypodermis, and hair follicles

2. _____ helps prevent dehydration of deep body tissues:
   (a) Keratin
   (b) Melanin
   (c) Glycogen
   (d) Sensory nerve basket
   (e) Ultraviolet light

3. An isotonic solution:
   (a) Has a concentration of about 0.9% NaCl dissolved in 99.1% H_2O
   (b) Seldom exists within the natural body fluids
   (c) Results in a net osmosis of water from the blood plasma into the RBCs
   (d) Is too watery to promote homeostasis
   (e) Is the exact opposite of a hypotonic solution

4. When red blood cells crenate, it means that they:
   (a) Abnormally swell up with fluid
   (b) Divide extensively by mitosis
   (c) Clump together to create blood clots
   (d) Shrink and wrinkle up
   (e) Blow up into tiny pieces!

5. Main nerve endings in the dermis that detect a rise in body temperature:
   (a) Thermoreceptors
   (b) Chemoreceptors
   (c) Hair follicle bulbs
   (d) Layered strata
   (e) Pores

6. Dense (compact) bone tissue consists primarily of:
   (a) Spongy bone
   (b) Yellow marrow
   (c) Haversian systems (osteons)
   (d) Osteoblasts
   (e) Embryonic connective tissue

7. Involves the depositing of calcium phosphate crystals:
   (a) Hematopoiesis
   (b) Bone fracture
   (c) Ossification
   (d) Body temperature homeostasis
   (e) Neuromuscular junction activities

8. Blood [Ca^{++}] measured in units of mg/dL blood is an example of:
   (a) An anatomical parameter
   (b) A functional parameter
   (c) Upper normal limit
   (d) A physiological parameter
   (e) The normal range of a parameter

9. If blood calcium ion concentration rose far beyond its upper normal limit, then the person would experience the symptoms of:
    (a) Normocalcemia
    (b) Hypertension
    (c) Normoglycemia
    (d) Hypocalcemia
    (e) Hypercalcemia

10. Osteoclasts primarily function in this process:
    (a) Extraction of sodium ions
    (b) Resorption of bone matrix
    (c) Remodeling of the skin
    (d) Bone formation
    (e) Oxygen transport

11. Usually provides evidence of physiological hypertrophy:
    (a) Shrinkage of a skeletal muscle down to half its normal size
    (b) Complete fracture of a long bone into several pieces
    (c) Too much sugar in the diet
    (d) Remodeling of bones to become thicker and stronger
    (e) Remodeling of bones to become thinner and weaker

12. "The erythrocyte is an anucleate, biconcave disc"; therefore:
    (a) RBCs have no nucleus but are shaped like an hourglass
    (b) Leukocytes must have exactly the same characteristics
    (c) RBCs have many nuclei and are shaped like globes
    (d) Each erythrocyte has the form of a doughnut
    (e) No detailed information about its structure is available

13. The most common type of leukocyte:
    (a) Platelet
    (b) Neutrophil
    (c) Megakaryoblast
    (d) Myeloblast
    (e) Thrombocyte

14. The fulcrum in any bone–muscle lever system:
    (a) Red bone marrow
    (b) Tendon
    (c) Movable joint
    (d) Muscle organ
    (e) Ligament

15. The major central portion of the neuron:
    (a) Soma
    (b) Dendrite
    (c) Axon
    (d) Axon terminals
    (e) Cytoplasm

16. The two general types of cells in the body that can have action potentials:
    (a) Neurons and liver cells
    (b) Bone cells and muscle fibers
    (c) Cartilage cells and lacunae
    (d) Muscle fibers and nerve cells
    (e) Muscle cells and epithelial cells

17. Stimulate the muscle fiber across the motor end plate:
    (a) Lateral sacs
    (b) Excitatory neurotransmitter molecules
    (c) Glucose molecules
    (d) Inhibitory neurotransmitter molecules
    (e) I bands

18. Portion of a myofibril where contraction actually occurs:
    (a) T-tubules
    (b) Sarcoplasmic reticulum
    (c) Sarcomere
    (d) Sarcolemma
    (e) Lateral sacs

19. Molecules arranged into a double helix attached at one end to a Z-line:
    (a) Myosin
    (b) Creatine
    (c) Creatine phosphate
    (d) Myosin ATPase
    (e) Actin

20. Provides the tilting action to help myofilaments slide:
    (a) Myosin cross-bridges
    (b) Muscle ADP
    (c) Troponin
    (d) Tropomyosin
    (e) ACh

21. A quick, jerky contraction of one or more motor units whose stimulus threshold has been reached:
    (a) Tetanus
    (b) Twitch
    (c) Subthreshold contraction
    (d) Summation
    (e) Relaxation response

22. _____ can be conveniently summarized as E–C–R:
    (a) Neuromuscular junction
    (b) Diffusion of Ca^{++} ions
    (c) All-or-None Law
    (d) Muscle activity cycle
    (e) Subthreshold excitation

23. Smoothly handing a person a piece of paper would likely involve:
    (a) Single twitches
    (b) Forced expiration
    (c) Summation
    (d) Fused tetanus
    (e) Relaxation

24. If Alfonso, the talking parrot, who is educated in muscle physiology, says, "LO," he really means:
    (a) Fibers with quick contractions
    (b) Cells which mainly rely upon glycolysis alone for their ATP production
    (c) Muscle fibers with a high density or crowding of mitochondria
    (d) "*Low* ceiling, *overhead*!"
    (e) Cells which chiefly depend upon the Krebs cycle and electron transport system for their ATP synthesis

25. The biochemical muscle fiber type probably most important in the leg muscles of a tramp who is determined to walk across America:
    (a) HOFT
    (b) LOST
    (c) LOFT
    (d) HOST
    (e) TOAST

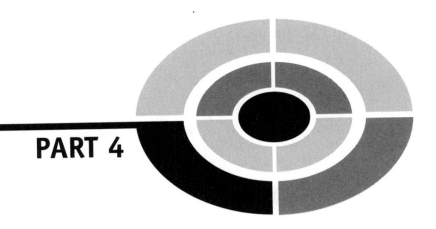

PART 4

The Physiology of Nerves and Glands

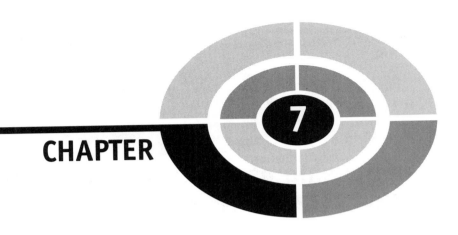

"What Happens When We Step on a Nail?": The Physiology of Nerves and Reflexes

Having discussed the physiology of muscles, bones, and joints, it is now time to delve more deeply into the "spark plugs" and the "electrical wires" that stimulate these body structures to move. Chapter 6 did bring up the concept of the neuromuscular junction (motor end-plate). And the basic anatomy of the *individual* neuron and its nerve fibers (axon and dendrites) was provided. If the individual neuron is a "spark plug," then the functional connections between the individual neurons and the associated nerves are the "electrical wires" that communicate with one another and carry information for long distances through the body. It is the main task of this chapter to explain how such communication of body information takes place.

Relationship of the Two Major Divisions of the Nervous System

There are two basic anatomical divisions of the nervous system: the *Central Nervous System* or *CNS*, and the *Peripheral* (per-**IF**-er-al) *Nervous System* or *PNS*. (Consult Figure 7.1.) The Central Nervous System (CNS) is the portion of the nervous system that is "centrally" located, along the body midline. It consists of the brain or *encephalon* (en-**SEF**-ah-**lahn**) plus the *spinal cord*.

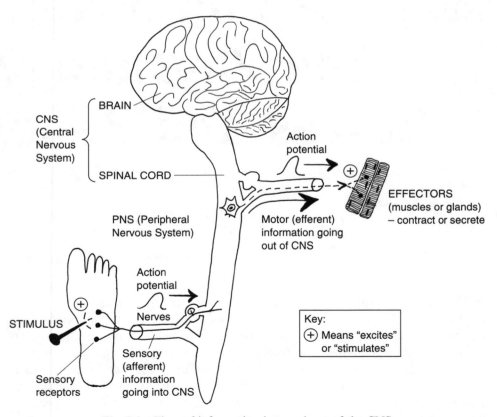

Fig. 7.1 Flow of information into and out of the CNS.

The Peripheral Nervous System (PNS), in contrast, is the portion of the nervous system that is located outside the brain and spinal cord, within the body "edge"or *periphery* (per-**IF**-er-ee). The PNS largely consists of nerves and *sensory receptors* that lie outside the CNS.

To simply summarize, we have:

THE	=	CENTRAL	+	PERIPHERAL
NERVOUS		NERVOUS SYSTEM		NERVOUS SYSTEM
SYSTEM	=	(CNS)		(PNS)
		(Brain + Spinal cord)	+	(Nerves + Sensory receptors)

Organ System 1

NERVES CARRY INFORMATION THROUGH THE PNS

Remember (Chapter 6) that nerve fibers are just long branches of a neuron's cytoplasm. They are either axons or dendrites. A *nerve* is a slender, white, thread-like collection of nerve fibers that carries information through the PNS. The information (in the form of action potential waves) travels through those parts of the body lying outside of the CNS (brain and spinal cord). Such parts of the body edge or periphery would include the arms and legs, for example.

SENSORY NERVE FIBERS BRING INFORMATION *TOWARD* THE CNS

Sensory receptors are modified nerve endings that are sensitive to particular stimuli. *Free nerve endings* in the dermis of the skin, for instance, are sensory receptors that are especially sensitive to pain and temperature stimuli. When a person steps upon a sharp nail, for instance, the free nerve endings in the sole of the foot are activated by this stimulus.

Action potential waves are created and travel over *sensory or afferent* (**AF**-fer-ent) *nerve fibers*, "toward" (*af-*) the CNS. They provide the CNS with *sensory* information about the stimulus (in this case, skin contact with a sharp nail). Thus:

SENSORY	→	SENSORY	→	CNS (receives
RECEPTORS (excited		(AFFERENT)		sensory
by a stimulus)		NERVE FIBERS		information)

Tissue 1

MOTOR NERVE FIBERS TAKE INFORMATION *AWAY FROM* THE CNS

There is an *opposite* direction of information flow (action potential travel), too. There are thousands of motor neurons present within the CNS. When some of these motor neurons "decide" to fire an action potential, these waves eventually travel over *motor or efferent* (**EE**-fer-**ent**) *nerve fibers*. They are called efferent nerve fibers, because they "carry" (*fer*) motor (movement-related) information "away from" (*ef-*) the CNS.

"Where do these motor or efferent nerve fibers eventually wind up, Professor Joe?" They go out to particular *effectors* (e-**FEK**-ters). An effector is literally "something that" (*-or*) carries out a particular response, thereby having some "effect" upon the body. Specifically, the effectors in the human body are of two basic types: they are either muscles that contract (causing the effect of body movement), or *glands* that *secrete* (releasing some useful substance).

Summarizing the above, we have:

Tissue 2

| MOTOR NEURONS ("decide" to fire action potentials) | → | MOTOR (EFFERENT) NERVE FIBERS | → | BODY EFFECTORS RESPOND (muscles contract, or glands secrete) |

The Synapse: Where Two Neurons "Clasp Hands"

If we are going to talk about long nerve fibers carrying motor or sensory information long distances through the body, we also have to talk about what happens when the fibers finally reach their destination! Usually we are dealing with neuron axons, which have those branching axon terminals at their tips. In the case of the axon terminals ending upon muscle effectors, the area of near-contact, you will remember, is the neuromuscular junction (motor end-plate). (Consult Figure 7.2, A.)

"But what happens if the axon terminals end upon part of another *neuron*, instead of a *muscle*, Professor Joe?" The resulting meeting place is called a *synapse* (**SIN**-aps). A synapse is a "clasping together" of two neurons, as between the axon terminals of one neuron and the dendrite or cell body of

Tissue 1

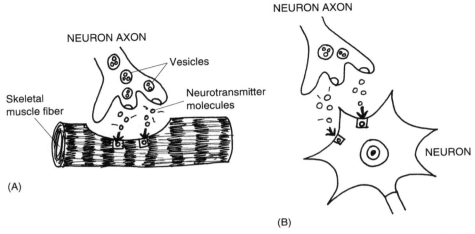

Fig. 7.2   The neuromuscular junction versus the synapse. (A) The neuromuscular junction
(motor-end-plate). (B) The synapse.

another. As is evident from Figure 7.2 (B), however, the synapse is a narrow,
fluid-filled gap where the axon terminal of one neuron *almost*, but not quite,
touches the cell body or dendrite of another neuron.

CHEMICAL COMMUNICATION VIA NEUROTRANSMITTERS

"If the two neurons don't quite touch, then how does one influence the
other?" The influence is accomplished in the same general way we have
seen for the neuromuscular junction – by chemical communication using
neurotransmitter molecules.

When the action potential reaches the end of the axon terminals of the first
neuron, it causes a number of the little, bladder-like, *synaptic* (sih-**NAP**-tik)
vesicles to rupture. Their content of neurotransmitter molecules is then
released into the fluid-filled space between the two neurons. (The neuron
positioned "after" the "synapse" is technically called the *postsynaptic* [**post-**
sih-**NAP**-tik] *neuron.*)

The released chemical diffuses across the synapse, and attaches to transmitter binding sites on the cell membrane of the postsynaptic neuron. If an
excitatory neurotransmitter (which is often ACh, acetylcholine) is released,
then its molecules excite the postsynaptic neuron, thereby making it more
likely to fire an action potential of its own. Communication across long
distances in the body, therefore, is often accomplished by a long chain of
neurons that release excitatory neurotransmitter molecules into the synapses

Cell 1

between them. An action potential in one neuron is soon followed by the creation of an action potential in the second neuron, which excites the third neuron in sequence, and so on.

Inhibitory neurotransmitters are a real "turn-off"!

"But if all synapses involve the release of only excitatory neurotransmitters, is there any way to turn the whole sequence *off*?", Baby Heinie asks, showing considerable intelligence for a little kid. Yes, there is. The flow of excitation down a long chain of neurons can be interrupted when an *inhibitory neuron* is excited. An inhibitory neuron is a neuron that releases *inhibitory neurotransmitter molecules* from its ruptured synaptic vesicles. Two chemicals that often act as inhibitory transmitters are called *GABA* and *glycine* (**GLEYE**-seen).

When either GABA or glycine is released, it diffuses across the synapse and inhibits the postsynaptic neuron. Thus, the postsynaptic neuron is less likely to fire off an action potential of its own, and the chain of neuron excitation is effectively turned off.

Neuron Axons: Resting Polarized, or Active Depolarized?

Cell 2

If the neuron cell body is a "spark plug," then the neuron axon could be considered a "spark plug wire." Let us take a really close look at the neuron soma (cell body) and an area called the *axon hillock* (**HILL**-ahk) (see Figure 7.3). A hillock is a "little hill." Therefore, the axon hillock is a little, raised, hill-like bump located near the base of the neuron axon. The axon hillock is of special functional importance, because it is the firing zone or trigger zone in the neuron where the action potential usually begins.

Cell 1

THE RESTING POLARIZED STATE

When a neuron is at "rest," electrochemically or ionically speaking, it is not conducting an action potential (Figure 7.3, A). The cell membrane of the neuron has an *electrical dipole* (**DIE**-pohl) or "double-pole" distributed across it. [**Study suggestion:** Visualize the positive (+) and negative (−) poles or ends of a car battery.]

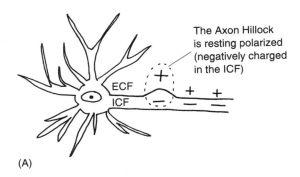

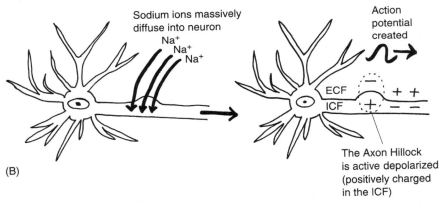

Fig. 7.3.   The neuron axon in its resting versus depolarized states. (A) The resting polarized state of a neuron. (B) The excited neuron: an active, depolarized state.

The intracellular fluid (ICF) within the neuron is negatively charged (that is, a negative pole) compared to the extracellular fluid (ECF) located outside of the neuron. Conversely, the ECF is positively charged (that is, a positive pole) compared to the ICF. In summary, there is a dipole across the neuron's plasma membrane: the *positive* (+) *pole* is on the *outside* of the neuron, while the *negative* (−) *pole* is on the *inside* of the neuron. This resting, non-conducting condition of the neuron is formally called the *resting polarized* (**POH**-lar-**eyezd**) *state*.

THE EXCITED NEURON: THE ACTIVE DEPOLARIZED STATE

When the neuron is excited, because of the binding of acetylcholine or some other excitatory neurotransmitter to it, the *sodium-ion channels* suddenly open within the axon hillock (Figure 7.3, B). There is a sudden great increase in the Na$^+$ ion permeability of the plasma membrane. Hence, there is a massive net diffusion of sodium ions from the ECF and into the ICF of the neuron.

With so many new, positively charged particles (Na$^+$ ions) now within the ICF, an *active depolarized* (**dee-POH**-lar-eyezd) *state* is created. Since *de-* means "away from," the word depolarized means "away from being (resting) polarized." In other words, the resting polarized state across the membrane (net positive charge outside the neuron, but net negative charge inside the neuron) has been *reversed*. The result is the creation of an action potential, which represents a traveling wave of depolarization (net positive charge inside of the axon, along with net negative charge outside of the axon). [**Study suggestion:** Picture the action potential as a + and − wave, with the + charge on the *inside* of the axon and the − charge on the *outside* of the axon.]

During this active depolarized state of the neuron cell membrane, an action potential is being actively conducted from its site of origin at the axon hillock, all the way down the axon.

Organelle 1

Neuron Axons: Naked, or Myelinated?

Once it is created, *how* an action potential travels largely depends upon whether the axon is *myelinated* (**MY**-el-in-**ay**-ted) or naked. A myelinated axon is one that is covered by a *myelin* (**MY**-eh-lin) *sheath*. Myelin is a soft, white, fatty insulating material found on the axons of certain neurons. A naked axon, in contrast, lacks a myelin sheath.

IS IT *GRAY*, OR IS IT *WHITE*? IT ALL DEPENDS UPON MYELIN!

Thousands of neuron cell bodies and dendrites, which lack any covering of myelin, make up the *gray matter* of the brain and spinal cord. Conversely, myelinated axons, with their white myelin sheath, make up the *white matter* of the brain and spinal cord. Some axons, however, are naked (lacking myelin).

LOCAL CIRCUIT CURRENTS, OR SALTATORY CONDUCTION? IT ALL DEPENDS UPON MYELIN!

Figure 7.4 suggests the two different ways that an action potential can be conducted down a neuron axon:

1. Naked axons: local circuit currents of ion flow. Naked axons can be compared to a bare copper wire. Would you want to grab a bare copper wire of a toaster, if it was plugged into a socket? Of course, you wouldn't! Because if you did grab the bare wire, you would get a nasty electrical shock! This is dangerous, because the naked copper wire lacks any external sheath of insulation.

In the naked copper wire, flowing electrons (in electricity) are free to enter and leave the wire all along its length. Similarly, in a naked neuron axon, the

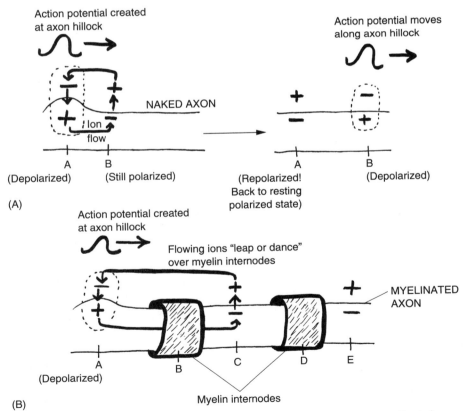

Fig. 7.4 Conduction of the action potential in naked versus myelinated axons. (A) Conduction along a naked axon: local circuit currents of ion flow. (B) Conduction along a myelinated axon: saltatory conduction.

action potential is conducted by *local circuit currents of ion flow*. It is best to think of these as small loops of flowing ions, moving into and out of the axon "locally" – at points immediately neighboring one another. There is a net inward flow of anions (such as Cl^-) toward the positively charged pole within the axon, followed by a net outward flow of cations (such as Na^+) toward the negatively charged pole just outside the axon. This repeating cycle of inward and outward ion flow, therefore, creates multiple local loops or circuits. As a result, the action potential is *propagated* (**PRAHP**-ah-**gayt**-ed) or "reproduced" again and again, until it finally reaches the tip of the axon terminals.

2. Myelinated axons: saltatory conduction of the action potential. In myelinated neurons, the axon is partially coated with a myelin sheath. We might compare this situation with a toaster whose copper wire (naked axon) is partially covered by an insulation of white rubber (myelin) (see Figure 7.4, B).

Suppose that our hero, Baby Heinie, being in a destructive mood, took a knife from the kitchen drawer and cut knicks out of the white rubber insulation that originally completely covered the toaster's plug-in cord. The places where the rotten kid cut down to bare copper wire (naked axon) we will call the *nodes of Ranvier* (rahn-vee-**AY**). Each of these nodes is a "knot" or narrowed region along the myelin sheath.

A series of *myelin internodes* (**IN**-ter-**nohds**) are found in the spaces "between" (*inter-*) the "knots" (*nodes*) of Ranvier. Each myelin internode is a roughly cylinder-shaped bead of myelin that occurs between two nodes of Ranvier. The overall effect is somewhat like a series of barrel-shaped Indian beads along a string necklace.

Since we have used a toaster cord as our main model, we will return to it. Of course, Dad wouldn't want to touch the toaster cord at any of its nodes of Ranvier! He would certainly receive an electrical shock from the flow of current leaking out of the cord at these naked points! A somewhat parallel situation exists for the real myelinated nerve fiber. There is a *saltatory* (**SAL**-tah-**toh**-ree) *conduction* of the action potential.

Saltatory conduction is a type of conduction or carrying of the action potential by a process of "leaping." Say that there are five adjacent points (A, B, C, D, and E) along a neuron axon. Suppose that points A, C, and E represent nodes of Ranvier (where ions can flow into and out of the axon). But points B and D are at myelin internodes (points of insulation where ions cannot enter or leave the axon). What happens, therefore, when an action potential begins at point A (the axon hillock)? The action potential "leaps" over and under the myelin internode at point B, so that the action potential is regenerated at point C. Again, there is a saltatory (leaping) flow of ions over and under the myelin internode at point D, so the action potential is re-created at point E. By this leaping process that bypasses myelin internodes of insulation, then, the action potential is moved or propagated all the way down to the end of the axon terminals.

Dad Steps on a Nail: Information Flow Through Nerves

"Okay, Professor Joe. We know we have this information, in the form of action potentials or ionic waves, traveling long distances over neuron axons. And with the help of synapses, this information can be directed into certain areas. What are some of these areas?"

NERVES: ANATOMICAL CONNECTIONS TO THE CNS

Well, we know that information mainly flows into and out of the CNS. So that is the main area where action potential information is being channeled. And as we have already pointed out, it largely consists of sensory information that travels toward the CNS over afferent nerve fibers, as well as motor information that travels away from the CNS over efferent nerve fibers. We just haven't gotten into a discussion of the related anatomy and physiology of the nerves carrying these nerve fibers.

Peripheral nerves are the farthest ones out

The PNS (Peripheral Nervous System) has been identified as the portion of the nervous system that is located in the body "edge" or periphery. Not surprisingly, the *peripheral nerves* are individual nerves traveling the farthest out in the body "edge."

The peripheral nerves are usually named for either the bones by which they pass, or for the body areas through which they pass. Let us consider a few specific examples. The *radial* (**RAY**-dee-al) *nerve*, for instance, travels along the *radius* (**RAY**-dee-us), a long "rod" (*radi*) of a bone on the thumb side of the forearm. And the *brachial* (**BRAY**-kee-al) *nerve* passes through the "upper arm," which is called the *brachium* (**BRAY**-kee-um) in Latin.

In the lower extremity, we have a number of major peripheral nerves. We will start with the most inferior nerves, then follow them progressively upward. The pathway begins with the *medial* (**ME**-dee-al) *plantar* (**PLAN**-tar) *nerve*, which serves the skin covering the "middle" (medial) 2/3 of the *plantar* (**PLAN**-tar) surface of the foot. The word plantar literally "pertains to the sole (*plant*) of the foot."

Suppose that Baby Heinie goes to bed, but leaves a sharp nail on the kitchen floor. Dad comes down after dark, and you know what happens! He steps on the nail! Dad steps on the nail with the middle of the sole of his right foot. Thus, the nailhead serves as a stimulus for free nerve endings (pain receptors) in the sole. These pain receptors are connected to sensory (afferent) nerve fibers that pass through the medial plantar nerve (Figure 7.5).

Action potentials travel from the pain receptors in the sole, through the medial plantar nerve, and then up into a more superior peripheral nerve – the *tibial* (**TIB**-ee-al) *nerve*. The tibial nerve ascends (travels up) through the lower leg, just posterior to the *tibia* (**TIB**-ee-ah), which is the "shinbone."

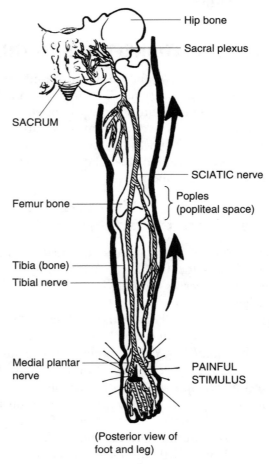

Hip bone

Sacral plexus

SACRUM

SCIATIC nerve

Poples (popliteal space)

Femur bone

Tibia (bone)

Tibial nerve

Medial plantar nerve

PAINFUL STIMULUS

(Posterior view of foot and leg)

Fig. 7.5 Peripheral nerves travel up to a spinal nerve plexus.

The tibial nerve keeps ascending, snaking upward through the *poples* (**PAHP**-lees), which is also called the *popliteal* (**pahp**-lit-**EE**-al) *space*. In Latin, the word poples means "ham of the knee" or the "hollow of the knee." [**Study suggestion:** Using your fingertips, probe the poples (popliteal space) as the indented hollow region in the back of your knee. Imagine the tibial nerve passing through it, as you carefully look at Figure 7.5.]

Next in sequence is the *sciatic* (sigh-**AT**-ik) *nerve*. The word sciatic comes from the Greek for "hip joint," and from *ischium* (**IS**-kee-um), which means "hip." Sciatic, therefore "pertains to the hip, hip joint, or ischium." The ischium, of course, is not the entire bony hip. Rather, it is the most inferior and posterior (lowest and most behind) portion of the hip bone.

Although it is named for the hip, the majority of the sciatic nerve runs from the poples (knee dent), up along the back of the *femur* (**FEE**-mur), or "thigh" bone. For a short distance above the femur, the sciatic nerve (as its name indicates) rises just posterior to the ischium region of the lower hip bone. Most interestingly, the sciatic nerve is the largest nerve in the entire body!

Some peripheral nerves run into a spinal nerve plexus

Despite the fact that it is the biggest nerve in the body, the sciatic nerve doesn't keep running along forever. Instead, it merges into the *sacral* (**SAY**-kral) *plexus* (**PLEKS**-us). A plexus, in general, is some kind of "braid" or network. There are several *spinal nerve plexuses* – braids or branching networks of *spinal nerves* supplying a particular area of the body. (A *spinal nerve* is a *nerve* that connects directly onto the *body* or main mass of the *spinal* cord.)

The sacral plexus, in particular, is a branching network of spinal nerves that supply the area around the *sacrum* (**SAY**-crumb). The sacrum is literally the "sacred" (*sacr*) bone. Because of its unique, pointed arrowhead shape, the sacrum was often considered "sacred" and used in magical ceremonies!

The sacral plexus is actually a branching network of *spinal nerves L4* and *L5* – the *fourth and fifth lumbar* (**LUM**-bar) *spinal nerves* – plus *spinal nerves S1 through S5* – the *first through fifth sacral spinal nerves*.

SPINAL NERVE PLEXUS TO SPINAL CORD

From the sacral plexus, let us follow one of the spinal nerves (Figure 7.6). Each spinal nerve splits into two *spinal nerve roots* as it nears the body of the spinal cord. There is a *dorsal root* more in "back" (*dors*), and a *ventral root*

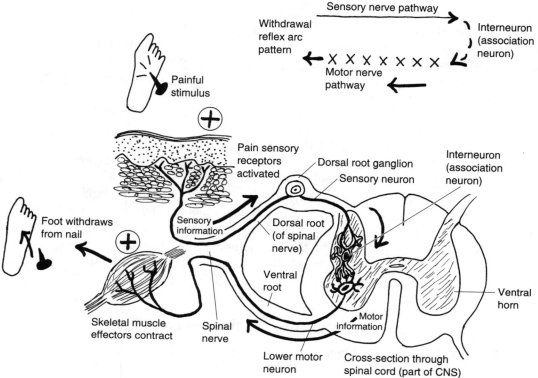

Fig. 7.6. Dad steps on a nail: a withdrawal reflex arc.

more on the front or "belly" (*ventr*) side of each spinal nerve. The dorsal root and ventral root (like the roots of a tree) serve to anchor each spinal nerve into the spinal cord. All of the spinal nerves, as well as their dorsal and ventral roots, of course, are considered part of the Peripheral Nervous System.

A Withdrawal Reflex Arc

What we have been tracing, ever since Dad stepped on that sharp nail on the kitchen floor with the sole of his right foot, is a type of *withdrawal reflex arc*. A reflex, in general, is some type of automatic, involuntary response to a particular stimulus. In this case, it is the automatic response to a painful stimulus delivered to the bottom of the right foot.

A *withdrawal reflex* is an automatic withdrawal of a body part from some harmful or annoying stimulus (such as a sharp nail pressing into the bottom of the foot). A *reflex arc* is a curved, arc or arch-like arrangement of the components of a reflex. What we are seeing in Figure 7.6, then, is a curved, arc-like arrangement of the three major types of neurons, and their nerve fibers, that are involved in automatically pulling Dad's right foot away from the painful nail stimulus.

Organ 1

THE SENSORY NEURON AND DORSAL ROOT

The first neuron to be activated in the withdrawal reflex arc is the *sensory neuron*. The dendrite of the sensory neuron has its peripheral end modified to form the sensory receptors for pain. In fact, everything back in Figure 7.5, from the peripheral nerves all the way up into the sacral plexus, was just the dendrite of a sensory neuron!

"You mean, then, Professor," Baby Heinie interjects, "that this whole thing has been just *one* dendrite?" Well, we have been tracing the ascending pathway followed by just one dendrite, but of many sensory neurons, running side by side, and serving the sole of the right foot. And, as you can plainly see, these dendrites can be quite long, extending from the foot all the way up into the spinal cord!

The dendrite of the sensory neuron carries afferent (sensory) information about the painful stimulus toward the CNS. As it approaches the spinal cord, the dendrite of the sensory neuron goes up into the dorsal root of the spinal nerve. Observe in Figure 7.6 that there is a little "knot," called the *dorsal root ganglion* (**GANG**-lee-ahn), present on the dorsal root. A ganglion, in general, is a knot-like collection of gray matter (neuron cell bodies and dendrites) located *outside* the CNS (brain and spinal cord). The dorsal root gangion in particular, however, contains the cell bodies of the sensory neurons only.

From the dorsal root ganglion, the axon of the sensory neuron continues a bit farther, finally terminating in axon terminals within the *horns of gray matter* in the spinal cord.

THE INTERNEURON (ASSOCIATION NEURON)

If we straighten all of the sensory nerve pathways into a smooth line, they look like the first part of a curved arc, which has been placed on its side. "What about the short bend or curve in the arc?", Baby Heinie follows this up.

That part is made by the *interneuron* (**IN**-ter-**nur**-ahn), which is alternatively called an *association neuron*. An interneuron (association neuron) is a small, short neuron that forms a functional "association" or bridge "between" (inter-) two much larger neurons. In this case, the small inter-neuron forms a functional association or linkage between the sensory neuron (bringing information into the spinal cord about the painful stimulus) and the motor neuron, which carries information out of the spinal cord.

THE LOWER MOTOR NEURON

The final, bottom portion of the withdrawal reflex arc is a *lower motor neuron*. We have already discussed motor neurons and the motor end-plate. A *lower* motor neuron, however, is a neuron whose cell body lies in the *lower part* of the CNS, often in or near the spinal cord.

In our current example, the cell body of the lower motor neuron lies in the *ventral* (belly-side) *horn* of gray matter within the spinal cord. The axon of this lower motor neuron leaves the spinal cord and immediately enters the ventral root of a spinal nerve. The ventral root of a spinal nerve, therefore, is "pure," in the sense that it carries motor or efferent nerve fibers only.

The ventral root hooks into a spinal nerve, which is always "mixed." This means that the spinal nerve essentially acts as a two-way street. It carries both sensory (afferent) nerve fibers bringing information *in* about the painful stimulus, as well as motor (efferent) nerve fibers taking information *out*, and directing body movement.

THE FINAL WITHDRAWAL RESPONSE

As the axon of the motor neuron leaves the individual spinal nerve, it enters the sacral plexus. From here, the axon of the motor neuron becomes a motor (efferent) nerve fiber that basically travels in the reverse direction of the sensory (afferent) nerve fiber. But the final destination is different: the skeletal muscle effectors of the lower leg. These skeletal muscles are stimulated to contract, thereby withdrawing the sole of Poor Dad's right foot away from the painful nail stimulus.

Sensory Pathways and Ascending Nerve Tracts

"But when the withdrawal reflex is over, and Dad has pulled his foot away from the nail, wouldn't he feel pain – and probably experience anger?", Baby

Heinie somewhat guiltily whispers. "I don't see any place in Figures 7.5 or 7.6 where such pain and anger could be experienced."

So right you are, Baby Heinie! Dad would almost certainly experience *plantalgia* (plan-**TAL**-jee-ah), or "pain" (*-algia*) in the "sole of the foot" (*plant*). For this, the *sensory areas of the brain* would have to be involved. In technical terms, a *sensory homunculus* (hoh-**MUNG**-kyuh-lus) – a "little feeling man or dwarf" – would have to be activated within the *cerebral* (seh-**REE**-bral) *cortex* (**KOR**-teks).

As Figure 7.7 shows, such a sensory homunculus can be visualized within the *postcentral* (**POST**-sen-tral) *gyrus* (**JEYE**-rus). This is a raised "ring" (*gyr*) or fold of cerebral cortex – the thin "bark" of gray matter covering the surface of the cerebrum. It is called the postcentral gyrus because it is located just "after" (*post-*) an important groove, called the *central sulcus*

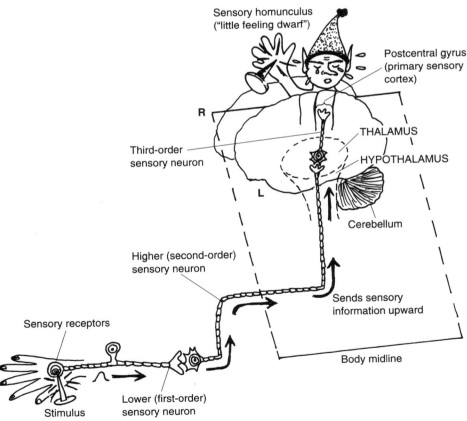

Fig. 7.7.   The spinothalamic tract ascends up to the feeling dwarf!

Tissue 3

(**SUL**-kus). Its other name is a more physiological one – the *primary sensory cortex*. It gets this name from the fact that general body senses, meaning the senses that can be experienced almost anywhere in the body (such as touch, pressure, pain, and temperature), are largely experienced in this postcentral gyrus or "primary" (basic) sensory cortex.

"But how do such general body sensations, such as pain, get all the way up there to the postcentral gyrus or primary sensory cortex? That's way up toward the top of the cerebrum, you know." They get there by means of *ascending or sensory nerve tracts*. A *nerve tract* is a collection of nerve fibers (axons and/or dendrites) that travels and stays within the CNS (brain and spinal cord). Ascending or sensory nerve tracts, consequently, are collections of nerve fibers that carry sensory information about particular stimuli, from lower levels of the CNS up to higher levels of the CNS.

In addition, ascending (sensory) nerve tracts usually *decussate* (dee-**KUS**-ate) to the opposite side of the body midline. Decussate literally means "to make an X." This is because sensory nerve fibers ascending from the right side of the body decussate or cross over to the left side of the body, and sensory nerve fibers from the left side ascend and cross over to the right side. As these ascending (sensory) fibers cross over the midline from opposite sides of the body, they decussate (trace an X) over each other.

THE SPINOTHALAMIC TRACT

One of the most important ascending (sensory) nerve tracts is called the *spinothalamic* (**spy**-noh-thah-**LAM**-ik) *tract*. The spinothalamic tract gets its name from the fact that it begins in the "spinal" cord and ascends to the "thalamus" (**THAL**-ah-**mus**). The thalamus is an egg-shaped "bedroom" (*thalam*) – an oval region nestled deep within the cerebrum. The thalamus is often described as a *sensory relay center*.

In the case of Figure 7.7, we have pain receptors in the hand (instead of the sole of the foot) being activated by a sharp nail. A *lower or first-order sensory neuron* begins things, because it is the nerve cell that has its dendrites modified to form sensory receptors. It travels all along through the peripheral nerves, plexuses, and other anatomical components of the sensory (afferent) pathway, all the way up into the dorsal root of a spinal nerve. Within the gray matter of the spinal cord, the spinothalamic tract begins.

Of course, there will be a withdrawal reflex arc occurring, causing the hand (like the foot) to be automatically withdrawn from the painful nail stimulus. But at the same time, a *higher or second-order sensory neuron* is excited. Its axon decussates (crosses over the midline) from the right side of the body to the left. The second-order sensory neuron has its axon terminal end within the thalamus. This is the official termination of the actual spinothalamic tract.

But the thalamus is called a sensory *relay* center, isn't it? The relaying action (like passing a baton during a relay race) is by a relatively short *third-order sensory neuron*, whose cell body lies within the thalamus. The axon terminals of this neuron "relay" the sensory message up into the postcentral gyrus of the cerebral cortex. It is here that the sensory homunculus (Dad's "little feeling dwarf") experiences the actual sensation of pain. Various association nerve fibers can also send the signal to nearby regions of the cerebrum, where expressions of anger and speech can originate.

Motor Pathways and Descending Nerve Tracts

Of course, movements of the body do not always have to be involuntary reflex actions (such as automatic withdrawal from a painful stimulus). Body movements can also be caused by a voluntary desire or will to move by the brain. This process involves *motor pathways*, which make use of *descending nerve tracts* (Figure 7.8).

A descending or motor nerve tract is a tract that carries motor (movement-related) information from higher levels of the CNS down to lower levels of the CNS. One very prominent example of such a descending (motor) nerve tract is the *corticospinal* (**kor**-tih-koh-**SPY**-nal) *tract*.

As its name reveals, the corticospinal tract begins in the cerebral cortex (*cortic*) and descends down to the spinal cord (*spinal*), after it decussates to the opposite side of the body. A key area to look at is the region around the central sulcus. As for the destination of the spinothalamic tract – the post-central gyrus – we again have a gyrus bordering the central sulcus. In this case, however, it is called the *precentral gyrus*. This is because the gyrus is located "before or in front of" (*pre-*) the central sulcus.

Tissue 4

And for the corticospinal tract, the cerebral cortex is just the beginning of the tract (not its final termination). *Upper motor neurons* decide to voluntarily move, say, the right hand. [**Study suggestion:** Speculate as to which precentral gyrus (right or left) will most likely make the decision to "bend" or flex the right hand, then check with Figure 7.8.]

After they decussate, the descending nerve fibers synapse onto a *lower motor neuron* in the spinal cord. Its axon goes out into the PNS via the usual anatomical sequence, into a peripheral nerve, and finally to a neuro-

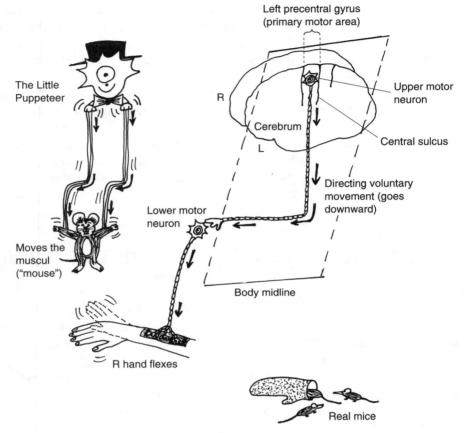

Fig. 7.8 The corticospinal tract and the Little Puppeteer.

muscular junction. A group of muscle fibers is thus voluntarily stimulated to contract, much like a bunch of excited "little mice" (*muscul*)!

Because of such skillful muscle action, one might even imagine a Little Puppeteer dwelling within the *primary motor area* of the *left precentral gyrus*, which is "pulling the strings" for the muscle effectors to contract and move.

Quiz

Refer to the text in this chapter if necessary. A good score is at least 8 correct answers out of these 10 questions. The answers are listed in the back of this book.

1. The part of the nervous system that contains the sensory receptors:
   (a) Nerves
   (b) PNS
   (c) Dorsal roots
   (d) Ganglia

2. Alternatively classified as afferent nerve fibers:
   (a) Motor nerve fibers
   (b) Axon hillocks
   (c) All dendrites
   (d) Sensory nerve fibers

3. Place where two neurons almost "clasp together":
   (a) Synapse
   (b) Motor end-plate
   (c) Neuromuscular junction
   (d) Myelin internode

4. Chemicals that make a postsynaptic neuron less likely to fire an action potential of its own:
   (a) ACh molecules
   (b) Excitatory neurotransmitters
   (c) Synaptic vesicles
   (d) GABA and glycine

5. The usual firing zone or trigger zone for an action potential:
   (a) Neuron soma
   (b) Myelin sheath
   (c) Node of Ranvier
   (d) Axon hillock

6. A stretch of a neuron cell membrane that is not currently conducting an action potential is accurately described as being:
   (a) Resting polarized
   (b) An electrical tripole
   (c) A neutral dipole
   (d) Active depolarized

7. Mode of action potential propagation that occurs along naked axons:
   (a) Saltatory conduction
   (b) Synaptic transmission
   (c) Local circuit currents
   (d) Ionic entrapment

8. The radial nerve is an example of a:
    (a) Cranial nerve
    (b) Peripheral nerve
    (c) Spinal nerve
    (d) Unmyelinated nerve

9. Consists of a branching network of spinal nerves L4 and L5, plus S1 through S5:
    (a) Conus medullaris
    (b) Nodes of Ranvier
    (c) Sacral plexus
    (d) Sciatic nerve

10. Sensory homunculus:
    (a) Site of origin of most sensory impulses
    (b) Place of final destination of most ascending nerve tracts
    (c) Association areas of the cerebellum
    (d) Demyelinated neuron axons

Body-Level Grids for Chapter 7

Several key body facts were tagged with numbered icons in the page margins of this chapter. Write a short summary of each of these key facts into a numbered cell or box within the appropriate *Body-Level Grid* that appears below.

Anatomy and **Biological Order** Fact Grids for Chapter 7:

A

CELL
Level

1

TISSUE
Level

1

ORGAN SYSTEM
Level

1

Physiology and   *Biological Order* Fact Grids for Chapter 7:

P

CELL
Level

1	2

TISSUE
Level

1	2
3	4

ORGAN
Level

1

Function and *Biological Order* Fact Grids for Chapter 7:

ORGANELLE
Level

1

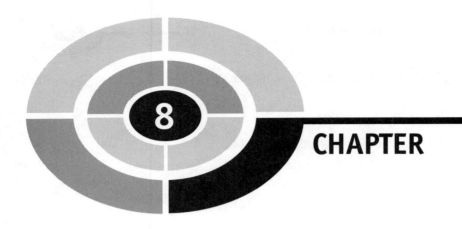

Glands: Physiology of the ''Ductless Acorns''

Chapter 7 compared neurons to spark plugs, and their axons to wires. "You must be going *nuts* to talk this way!", some of you may now be saying to yourselves. Well, let's do just that! Specifically, let's talk about some of the basic physiology of the *glands*, which literally are a type of nut – "acorns" in Latin.

Glands Versus *Glans*: They're All Some Kind of ''Acorn''!

In the world of *plant* anatomy, an acorn is the nut (dry fruit) of an oak tree. (A fruit is the female part of a plant, which contains the seeds.) It has a rounded body, with a pointed tip. The acorn is usually connected to a tree limb by a slender stalk. (Consult Figure 8.1, A.) And in plants, any rounded swelling (such as an acorn) can also be called a gland.

There are several glands in *human* anatomy that have very acorn-like shapes. These include the *pituitary* (pih-**TOO**-ih-**tair**-ee) *body* or *hypophysis* (high-**PAHF**-ih-sis), as well as the *pineal* (**PIN**-ee-al) *gland*. The hypophysis is literally "an undergrowth present" under the hypothalamus of the brain, and attached to it by the *pituitary* (pih-**TOO**-ih-**tair**-ee) *stalk*. Thus the

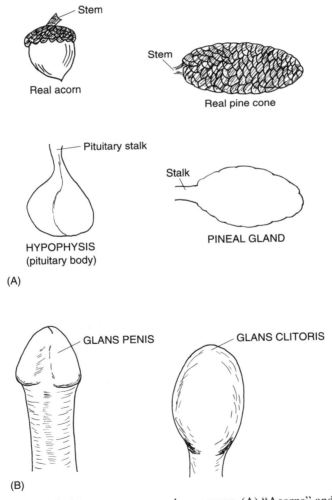

Fig. 8.1 Glands, glans: fruits, nuts, acorns, and sex organs. (A) "Acorns" and "pine cones." (B) Glans: "Acorn"-shaped heads.

hypophysis (pituitary body) does, indeed, resemble a grayish-colored acorn hooked onto the bottom of the brain by its stem!

A similar situation exists for the pineal gland, which is also associated with the brain. This gland is named for its close likeness to a "pine cone" – still another nut!

GLANS LET OUR SEX ORGANS GET "A HEAD"

A related word, *glans*, also means "acorn" in Latin. This type of "acorn," however, hardly grows on trees! "Why, Professor Joe?" Well, Baby Heinie, you are somewhat underaged, but you really need to be properly educated about your sex organs! "Yes, Professor Joe! If I don't get educated about my sex organs, how am I supposed to get a head in life?"

Precisely, Baby Heinie! A look at Figure 8.1 (B) shows two examples of glans that serve as rounded, acorn-shaped heads of several male and female sex organs. The *glans penis* (**PEA**-nis), for instance, is the acorn-like head of the male penis. Similarly, the *glans clitoris* (**KLIT**-er-is) is the acorn-shaped head of the female clitoris. The physiology of both these glans is closely involved with sexual stimulation and arousal.

In summary, the glands and heads of some of the sex organs in the human body are named for their anatomical shape – like an acorn – not for their physiology.

"What is the main physiology of glands, then, Professor?" Glands consist of one or more epithelial cells specialized for the function of *secretion* – the release of some useful substance.

Endocrine Glands Versus Exocrine Glands

Even though the entire general category of glands is named for its anatomy, the two major groups of glands are named for their differing physiology. Specifically, the two major types are named for the different *destinations* of their secretions (Figure 8.2).

Exocrine (**EKS**-oh-krin) *glands* are glands of "external" (*exo-*) "secretion" (*crin*) of a useful product into a *duct*, which carries the secretion to some body surface. A good example of exocrine glands are the sweat glands of the skin. The gland cells line the walls of the *sweat duct*, which carries the sweat up to the skin surface through a pore. Sweat can be considered a useful secretion, because it helps to cool the body and lower its temperature.

Organ 1

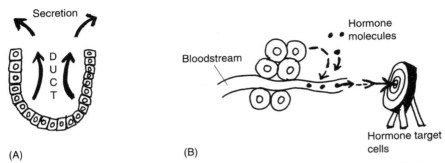

Fig. 8.2 Exocrine versus endocrine glands: two different modes of secretion. (A) Exocrine glands. (B) Endocrine glands.

Endocrine (**EN**-doh-krin) *glands* are glands of "internal" (*endo-*) "secretion" (*crin*) of a *hormone*, or chemical messenger, directly into the bloodstream, within the gland itself. Endocrine glands are therefore considered "ductless," since their hormone enters the bloodstream without passing through any duct. The hormone is then carried throughout the body.

The endocrine portion of the *pancreas* (**PAN**-kree-ahs), for instance, is called the *pancreatic* (**pan**-kree-**AT**-ik) *islets* (**I**-lets) or "little islands" of endocrine gland cells. Each pancreatic islet includes the *beta* (**BAY**-tuh) *cells*, which are well known for their function of secreting *insulin* (**IN**-suh-lin). Many of you already know that insulin is a major hormone that causes a decrease in the blood glucose concentration.

Organ 2

Important Principles of Hormone Action

The basic anatomy and descriptions of the various endocrine glands are provided in our companion volume, *ANATOMY DEMYSTIFIED*. This book, which emphasizes physiology, will focus upon important principles of hormone action. And insulin from the beta cells of the pancreas, being such a critical hormone, will often serve as our specific example.

HORMONE ACTION PRINCIPLE 1: HORMONES ARE "AROUSERS"

The word hormone literally means "to arouse, set into motion, or urge on." Hormones, therefore, do not create any new physiological or biological

Molecule 1

processes within our cells. They merely alter (often "arousing" or increasing) the rate at which these processes occur.

Glucose, for instance, enters most of our cells by the process of facilitated diffusion out of the bloodstream (Chapter 4). The protein carrier molecules within cell membranes are always functioning, but at a relatively slow rate. Glucose molecules still enter our body cells by facilitated diffusion, but this process does not occur at a significant rate. What insulin does, then, is "arouse" or stimulate the carrier proteins to work much faster. Consequently, glucose molecules enter the body cells at a much faster rate, and the blood glucose concentration falls.

HORMONE ACTION PRINCIPLE 2: ENDOCRINE GLANDS ARE LEAKY FAUCETS

A particular hormone (such as insulin) is always being secreted from its endocrine gland. This makes the gland behave much like a continuously dripping, leaky faucet. But the rate at which a hormone is secreted will change or vary with the strength of the gland's stimulus.

The major stimulus for the beta cells of the pancreas is a rise in the blood glucose concentration toward its upper normal limit. This would occur after eating a sugar-rich meal. The higher that the blood glucose concentration gets, the more strongly the beta cells are stimulated. And the more strongly the beta cells are stimulated, the greater will be their rate of insulin secretion (provided they are healthy, normal beta cells).

HORMONE ACTION PRINCIPLE 3: HORMONES OFTEN COMBINE WITH PLASMA PROTEINS

Most hormone molecules are not very soluble or "dissolvable" within the blood plasma (fluid portion of the blood). Instead, they combine with *plasma proteins*. The result is called a *hormone–protein complex*. The hormone–protein complex is quite soluble, allowing the hormone to travel long distances through the bloodstream. When the body destination of the hormone is reached, it breaks away from the plasma protein carrying it.

HORMONE ACTION PRINCIPLE 4:
"A LITTLE DAB WILL DO YOU"

Hormones are typically secreted in tiny *microgram* ("millionths of a gram") quantities, but they have dramatic physiological effects. Consider the hormone *epinephrine* (**ep**-ih-**NEF**-rin), which is more popularly known as *adrenaline* (ah-**DREN**-ah-lin). The physiological response to severe stress, often called the "fight-or-flight response," is carried out by the *sympathetic nerves* and the endocrine gland called the *adrenal* (ah-**DREE**-nal) *medulla* (meh-**DOO**-lah). Both the sympathetic nerves and the adrenal medulla are activated when the hypothalamus of the brain is prodded by stress (Figure 8.3).

Just a few additional micrograms of epinephrine, which is released from the adrenal medulla, triggers such severe "fight-or-flight" symptoms as

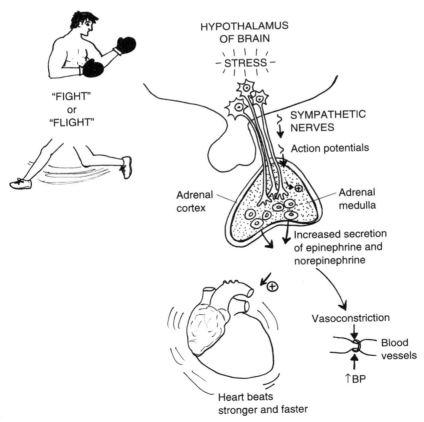

Fig. 8.3 Stress and the "fight-or-flight" response.

tachycardia (**tak**-ee-**KAR**-dee-ah) and powerful *vasoconstriction* (**vay**-soh-kahn-**STRIK**-shun). This is an extremely "swift" (*tachy-*) rate of beating of the "heart" (*cardi*), accompanied by a dramatic "narrowing" (*constriction*) of blood "vessels" (*vas*), which greatly raises the blood pressure.

HORMONE ACTION PRINCIPLE 5: THE BLOODSTREAM IS AN "UNSTOPPERED SINK"

"Baby Heinie, when was the very *first* time in your entire life that you were really mad and upset?," our Good Professor asks his somewhat difficult little companion.

"Oh, my Mom says my face was all red, and I was crying hysterically, the very moment I was born!"

Apparently, Baby Heinie (like practically all of us) experienced his first severe episode of the "fight-or-flight" response during the birth process. "In other words, Professor, this means that I was *born* mean, that I'm gonna *live* mean, and that I'm gonna *die* mean?"

"It does, Baby Heinie, if the bloodstream acts like a stopped sink for epinephrine and other stress hormones." What Professor Joe is trying to say is that if the few additional micrograms of epinephrine released during your very first episode of severe stress kept circulating and recirculating through your bloodstream again and again, then you would always stay mad. Simply speaking, you would never just be able to calm down and "get over it"!

This condition would exist if the bloodstream were like a stopped or plugged sink, never able to dispose of its current load of stress hormones. But we all know from practical experience that we usually *do* get over our severe episodes of stress, don't we?

Therefore, the bloodstream acts like an "unstoppered sink" for its current load of hormones. [**Study suggestion:** Visualize a sink with a faucet emptying into it. The faucet is the endocrine gland, the water is the hormone being released, and the sink represents the hormone's receiving area, the bloodstream. After the faucet has been running for a while, pull the plug or stopper out of the bottom of the sink. What happens to the contained "water" (current load of hormone molecules)?]

The unstoppered sink is reflected in the action of the liver and kidneys: hormone molecules are continuously inactivated by the liver, and then excreted into the urine by the kidneys. So this is the reason you usually don't stay mad for very long!

HORMONE ACTION PRINCIPLE 6: HORMONE MOLECULES ARE ARROWS FOR TARGETS

We can visualize each endocrine gland as a single faucet, and the water coming out of the faucet as its hormone. Now, since there are many endocrine glands, aren't there actually many different faucets, all dripping their water (hormones) into the same sink (the bloodstream)?

As the blood flows past the billions of cells in our body tissues, it is like a chemical soup, in that it contains dozens of different hormones. Quite obviously, a particular cell in the body cannot respond to all of these hormones!

"How does it choose which hormones to respond to, Professor?" It doesn't consciously "choose." Rather, we can view hormone molecules as arrows, which are aimed or directed at only certain targets. We can also change the analogy and view hormone molecules as keys, which can fit into the locks (*hormone receptor or binding sites*) of particular *target cells*.

We can thus state the principle of *Target Cell Specificity: A given hormone only affects certain specific target cells within the body – those having hormone receptor sites of the appropriate shape and size.* If the key (hormone molecule) floats by the lock (hormone receptor site on the surface of a cell), and fits into it, then the key can open the door (serve as an "arouser" or messenger for the cell).

Molecule 2

For the insulin molecule, its main target cells are the *adipocytes* (**AH**-dip-uh-**sights**) or "fat cells," *hepatocytes* (heh-**PAT**-uh-**sights**) or "liver" (*hepat*) "cells," and the muscle fibers (muscle cells). The insulin molecule is like a key that fits into the hormone receptor sites (keyholes) in the surface of the lock (plasma membrane) of adipocytes, hepatocytes, and muscle fibers.

Conversely, the neurons (nerve cells) are *non-targets* for insulin. They lack hormone receptor sites of appropriate shape and size, so insulin molecules cannot affect them. Glucose is the major fuel for neurons, and (like elsewhere) glucose enters the neurons by the process of facilitated diffusion. The process is fast enough to be significant, even without the help of insulin. We can therefore say that the facilitated diffusion of glucose into neurons is *insulin-independent*.

HORMONE ACTION PRINCIPLE 7: HORMONES ARE "FIRST MESSENGERS"

By definition, hormones are "arousers" that "set things into motion" after binding to the surface of their target cells. But the reason that there are more hormone molecules present in the bloodstream in the first place is because the gland that secreted them was prodded by some stimulus.

Consider the beta cells of the pancreas (Figure 8.4). They are stimulated to secrete more insulin, whenever the blood glucose concentration rises toward its upper normal limit (as after eating). Consequently, the added insulin molecules in the bloodstream act as "First Messengers," informing their target cells that the blood glucose concentration has risen.

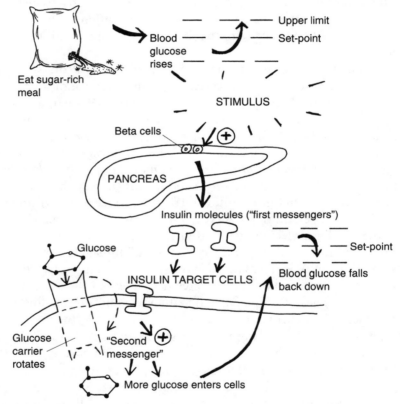

Fig. 8.4   Insulin and "Second Messengers" within its target cells.

HORMONE ACTION PRINCIPLE 8: HORMONES GET HELP FROM "SECOND MESSENGERS"

After hormones lodge into receptor sites on their target cell membranes, they have essentially given the target cells their "First Message." But an important problem still remains – that of actually carrying out a response to the "First Message."

Since they are usually too large to actually enter their target cells, many hormones just stay on the cell surface and rely upon the help of intracellular helper molecules. These helper molecules are called "Second Messengers." When insulin binds to the plasma membrane of one of its target cells, it activates the formation of more "Second Messenger" molecules within the target cell. The Second Messenger then actually carries out the appropriate change in target cell metabolism. In the case of insulin, the Second Messenger stimulates the activity of the carrier molecules involved in the facilitated diffusion of glucose into the target cell. (Review Figure 8.4.)

HORMONE ACTION PRINCIPLE 9: SECRETORY NEURONS ALSO PRODUCE HORMONES

The pituitary body (hypophysis) consists of two major subdivisions: an *anterior pituitary gland* located more in "front" (*anteri*), plus a *posterior pituitary gland* located "behind" (*posteri*). Both these front and back portions of the pituitary body are affected by hormones released from the hypothalamus, which lies just above them (Figure 8.5).

We already know that the hypothalamus (the region located below the thalamus) is considered part of the nervous system. And we know that the pituitary body contains two endocrine glands. Therefore, the hypothalamus–pituitary linkage makes up what can be called a *neuroendocrine* (**NUR**-oh-**en**-doh-krin) – "nervous system" (*neuro*) and "endocrine"–gland connection.

Organ System 1

While we have generally defined glands as collections of *epithelial* cells that secrete useful substances, there are also *secretory* (**SEE**-kreh-**tor**-ee) *neurons* within the hypothalamus that have very gland-like functions. Consider one specific type of hormone, called *antidiuretic* (**an**-tee-**die**-yuh-**RET**-ik) *hormone*, abbreviated as *ADH*. It is actually a product of the secretory neurons within the hypothalamus. Once secreted, however, antidiuretic hormone (ADH) passes into the bloodstream, which carries it down through the pituitary stalk.

ADH is temporarily stored within the cells of the posterior pituitary gland, then released into the general bloodstream. This hormone eventually circulates down to the kidneys, where it exerts its *antidiuretic effect*. Specifically, ADH promotes the *reabsorption* of more H_2O molecules from the *kidney tubules* (**TWO**-byools), the "tiny tubes" within the kidneys. As a result, it supplies a mechanism that acts "against" (*anti-*) "excessive urination" (*diuresis*). [**Study suggestion:** Look at the miniature "kidney faucet" model in

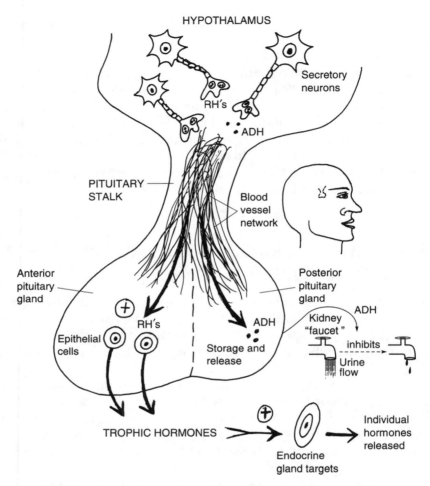

Fig. 8.5 The pituitary stalk: making a neuroendocrine connection.

Figure 8.5. How does the literal translation of the word antidiuretic help explain the turning off of the faucet from full blast to a slow drip?]

HORMONE ACTION PRINCIPLE 10: RELEASING HORMONES FROM THE HYPOTHALAMUS STIMULATE THE ANTERIOR PITUITARY TO SECRETE TROPHIC HORMONES

There is another neuroendocrine relationship between the secretory neurons in the hypothalamus and the anterior (rather than posterior) portion of the pituitary body. These secretory neurons produce *releasing hormones (RHs)*. These releasing hormones (like ADH) are secreted into the network of tiny blood vessels coursing down through the pituitary stalk. But they go down into the anterior pituitary gland (rather than the posterior pituitary gland). The releasing hormones get their name from their primary function. They literally stimulate the *release* of a variety of *trophic* (**TROHF**-ik) *hormones* from the epithelial cells of the anterior pituitary gland.

HORMONE ACTION PRINCIPLE 11: TROPHIC HORMONES FROM THE ANTERIOR PITUITARY "NOURISH" (STIMULATE) OTHER ENDOCRINE GLANDS TO SECRETE

"What are these trophic hormones you are talking about, Professor Joe? What does the word *trophic* mean in the first place?"

Trophic literally "pertains to" (-*ic*) "nourishment" (*troph*). But by "nourishment," we certainly *don't* mean that we are going to start handing out candy bars! A *trophic effect* is "nourishing" only in the sense of "stimulation."

As far as glands are concerned, the trophic hormones secreted by the anterior pituitary cells are "nourishing or stimulating" for *other endocrine glands*. The trophic hormones are unique in that their main targets are other endocrine glands, which are "nourished or stimulated" to increase the secretion of their own individual hormones. (Observe the little target in Figure 8.5.)

F

Molecule 3

Some specific endocrine gland targets and their hormones

Figure 8.6 shows pictures of the major endocrine gland targets being stimulated by various trophic hormones from the anterior pituitary. Take, for instance, the *thyroid* (**THIGH**-royd) *gland* in the front of the neck. The thyroid gland is named for its "resemblance" (-*oid*) to a broad "shield" (*thyr*). The thyroid is stimulated or "nourished" by *thyroid-stimulating hormone*, abbreviated as *TSH*. The thyroid gland is stimulated by TSH to increase the rate of secretion of its own individual hormone, *thyroxine* (**thigh-ROCKS**-in). Thyroxine, in turn, circulates throughout the bloodstream to affect most of the body cells. Thyroxine, for example, increases the *basal* (**BAY**-sal) *metabolic rate* or *BMR*, that is, the rate at which body cells burn calories during their metabolism under resting or "basal"

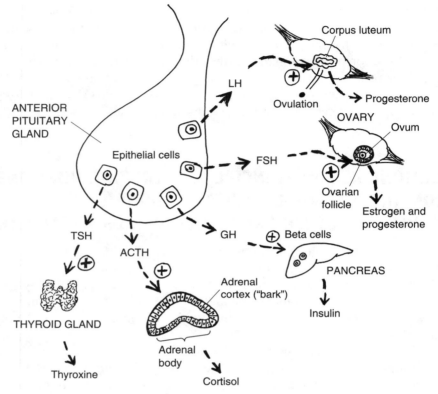

Fig. 8.6 Some major trophic hormones and their endocrine targets.

conditions. [**Study suggestion:** From its effect upon BMR, how would you expect thyroxine to influence oral body temperature? Why?]

The name of a second important trophic hormone is a real tongue-twister! Its name is *adrenocorticotrophic* (uh-**dree**-noh-**kor**-tuh-koh-**TROHF**-ik) *hormone*, abbreviated simply as *ACTH*. The *adrenocortico-* part of the hormone's name comes from its endocrine gland target – the *adrenal* (uh-**DREE**-nal) *cortex*. Just as the cerebral cortex literally forms a thin "bark" over the surface of the cerebrum, the adrenal cortex is an endocrine gland forming a thin "bark" (*cortex* or *cortico-*) over the surface of the *adrenal body*. This body is a curved, stocking cap-shaped structure that lies "toward" (*ad-*) the top of each "kidney" (*renal*).

The adrenal cortex secretes the hormone *cortisol* (**KOR**-tih-**sol**). Cortisol raises the blood glucose concentration whenever it is low, and it also acts to relieve the symptoms of tissue inflammation.

A third trophic hormone is called *growth hormone (GH)*. Growth hormone, as its name states, circulates to most of the body cells and stimulates their growth by promoting such processes as protein synthesis and cell division. But the specific target gland of its trophic influence are the beta cells of the pancreas. The beta cells, you may recall, secrete insulin. Insulin is absolutely critical for human survival, because it helps transport glucose out of the bloodstream and into the tissue cells, thereby providing them with their major energy source.

A fourth trophic hormone is *follicle-stimulating hormone (FSH)*. A follicle in general is any "little bag." Consider the case of the female *ovaries* (**OH**-var-**ees**), two white, "egg" (*ovari*)-shaped endocrine glands attached to both sides of the *uterus* (**YOU**-ter-us). Each ovary contains dozens of *ovarian* (oh-**VAIR**-ee-an) *follicles* – tiny bags or sacs that surround the developing *ovum* (**OH**-vum) or "egg" (*ov*) cell.

Under the stimulating effect of follicle-stimulating hormone (FSH), the ovarian follicles increase their secretion of the two hormones, *estrogen* (**ES**-troh-jen) and *progesterone* (proh-**JES**-ter-**own**). Estrogen stimulates the development of *secondary sex characteristics* in the female, such as a higher voice and softer skin. Progesterone prepares the female body for a possible pregnancy. As a part of this reproductive role, progesterone stimulates the lining of the uterus (womb) for potential imbedding of a fertilized ovum.

A fifth trophic hormone, *luteinizing* (**LEW**-tuh-**neye**-zing) *hormone* or *LH*, also acts upon the female ovary. It triggers a rupture of the mature ovarian follicle, thereby causing *ovulation* (**ahv**-you-**LAY**-shun) – the release of a "little egg" (*ovul*) into the abdominal cavity. But luteinizing hormone actually derives its name from a structure that appears after ovulation. This structure is called the *corpus* (**KOR**-pus) *luteum* (**LEW**-tee-um) or "yellow body." The corpus luteum is the tiny, yellowish-colored body that the ruptured ovarian follicle transforms into after the mature ovum has been released from it. LH thereby indirectly has a luteinizing or "yellowing" effect. It creates a yellowish body (corpus luteum) that secretes lots of progesterone (and some estrogen).

HORMONE ACTION PRINCIPLE 12: THE SECRETION OF HORMONES IS UNDER NEGATIVE FEEDBACK CONTROL

The final principle of hormone action is perhaps the most important of them all. The endocrine glands, much like the nervous system (and their

Molecule 4

neuroendocrine connection), are very much involved in the critical body functions of communication and control of various parameters within the internal environment.

Head Chief in the world of control and regulation, of course, is the negative feedback control system. Control of the internal environment, you may recall, was a major focus of our efforts back in Chapter 2. For the purposes of this chapter on glands, we can state the following general rule: *The secretion of hormones is under negative feedback control, such that high blood levels of a given hormone (or its predecessors) serves to "turn off" or inhibit further secretion of that hormone.*

This general principle will now be traced with regard to one of the specific anatomical parameters mentioned in Chapter 2. This was blood glucose concentration, measured in milligrams of glucose per 100 ml of blood.

Let us follow what happens after a person eats a sugar-rich meal, and how negative feedback is involved with regulation of both blood glucose level as well as the rate of insulin secretion. Figure 8.7 starts off with the major gland stimulus: a rise in blood glucose concentration toward its upper normal limit of 110 mg/100 ml blood. This stimulus is detected by the beta cells of the pancreas. They respond by dramatically increasing their rate of insulin secretion into the bloodstream. The insulin circulates to its target cells, where it indirectly triggers the Second Messengers to speed up the facilitated diffusion of glucose molecules out of the bloodstream and into the target cells.

This enhanced movement of glucose out of the bloodstream serves to decrease the blood glucose concentration back toward its set-point level of about 90 mg/100 ml blood. This final response of the control system has a negative feedback effect upon the initial stimulus (a rise in blood glucose toward its upper normal limit), acting to remove or "correct" it. The beta cells in the pancreas, having had their major stimulus removed, are now inhibited or "turned off" via negative feedback, so that they act like a gland faucet that is just barely "dripping" insulin!

Because insulin was present at fairly high levels, dropping the blood glucose significantly, it removed the stimulus for the beta cells. Hence, high blood levels of insulin eventually greatly inhibit further secretion of more insulin. This feedback control mechanism thus helps to prevent *insulin hyper-secretion* (**HIGH**-per-see-**KREE**-shun) – an "excessive or above-normal" (*hyper-*) amount of insulin release by the beta cells.

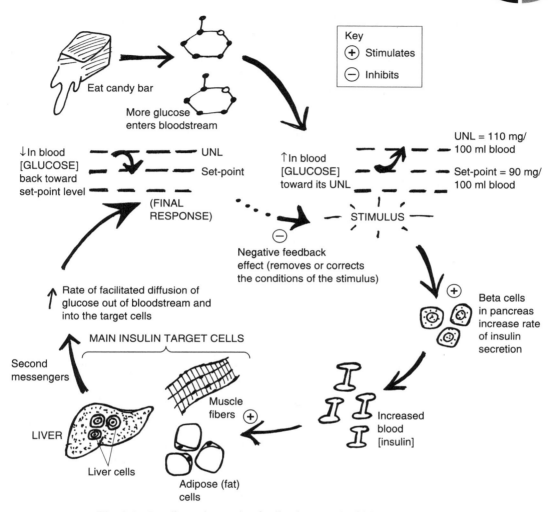

Fig. 8.7 Insulin and negative feedback control of blood glucose levels.

Quiz

Refer to the text in this chapter if necessary. A good score is at least 8 correct answers out of these 10 questions. The answers are listed in the back of this book.

1. An "acorn"-shaped gland attached to the base of the hypothalamus by a stalk:
   (a) Pineal
   (b) Pancreas
   (c) Pituitary body
   (d) Mammary

2. Sweat glands are of this type:
   (a) Exocrine
   (b) Internally secreting
   (c) Ductless
   (d) Mixed

3. Responsible for secreting insulin:
   (a) Ovaries
   (b) Adrenal cortex
   (c) Sympathetic nerves
   (d) Beta cells

4. Hormones are "arousers" in the sense that:
   (a) They create new body structures and functions that were not present earlier
   (b) Their influence always speeds up biological processes
   (c) They change the rate at which existing body processes occur
   (d) Removing them has no disturbing influence upon homeostasis

5. A hormone–protein complex is often essential, because:
   (a) No hormone can do a good job by itself
   (b) Many hormones are otherwise not very soluble in the bloodstream
   (c) All hormones are proteins anyway
   (d) They always carry opposite poles of charge

6. Importantly involved in the "fight-or-flight" response:
   (a) Parasympathetic nerves and sweat glands
   (b) The pineal gland
   (c) Sympathetic nerves and the adrenal medulla
   (d) Secretion of lots of progesterone into the bloodstream

7. Today, hormone molecule X is secreted into the bloodstream. One year from today, hormone molecule X will be:
   (a) At the very same concentration as it is today
   (b) Present at a much higher concentration than it is today
   (c) Converted into hormone Y, which is more powerful
   (d) Inactivated and excreted from the body

8. Neurons are non-targets for insulin, because:
    (a) They lack insulin receptor sites of appropriate shape and size
    (b) Nerve cells keep moving around too fast for insulin to catch them!
    (c) Enzymes keep pumping insulin out of them
    (d) Insulin has no targets!

9. Hormones are "First Messengers" in their:
    (a) Ability to activate "Pre-messengers"
    (b) Presence in higher concentrations within the bloodstream when their gland is being strongly stimulated
    (c) Tendency to clump blood cells
    (d) Frequence of occurrence whenever the body is injured

10. Provides strong evidence of a neuroendocrine relationship:
    (a) Presence of secretory neurons within the hypothalamus
    (b) Storage of glucose within hepatocytes
    (c) Association of the adrenal cortex with the adrenal medulla
    (d) Occurrence of anterior and posterior pituitary together in the hypophysis

Body-Level Grids for Chapter 8

Several key body facts were tagged with numbered icons in the page margins of this chapter. Write a short summary of each of these key facts into a numbered cell or box within the appropriate *Body-Level Grid* that appears below.

Physiology and *Biological Order* Fact Grids for Chapter 8:

ORGAN
Level

1	2

ORGAN SYSTEM
Level

1

Function and *Biological Order* Fact Grids for Chapter 8:

F

MOLECULE
Level

1	2

| 3 | 4 |
| | |

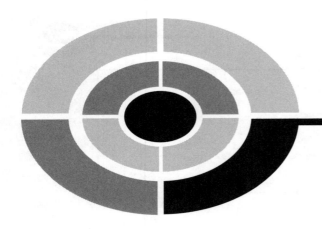

Test: Part 4

DO NOT REFER TO THE TEXT WHEN TAKING THIS TEST. A good score is at least 18 (out of 25 questions) correct. Answers are in the back of the book. It's best to have a friend check your score the first time, so you won't memorize the answers if you want to take the test again.

1. The spinal cord is considered part of the:
   - (a) CNS
   - (b) ANS
   - (c) PNS
   - (d) SNS
   - (e) SOS!

2. Afferent nerve fibers always travel:
   - (a) Away from the brain
   - (b) Through the spinal cord
   - (c) Toward the brain
   - (d) Parallel to the spinal cord
   - (e) In completely unpredictable patterns

3. The lower leg kicks out when the kneecap is hit with a hammer. The _____ carry out this action:
   (a) Muscular effectors
   (b) Hormone binding sites
   (c) Sensory receptors
   (d) Glandular afferents
   (e) Motor sensations

4. Efferent nerve fibers:
   (a) Travel from the heart, toward the spinal cord
   (b) Ascend and descend within the CNS
   (c) Carry information away from the body of the spinal cord
   (d) Are never involved in reflex arcs
   (e) Seldom supply any viscera

5. The narrow, saltwater-filled gap between the axon terminals of one neuron and the cell body and dendrites of another neuron:
   (a) Synapse
   (b) Node of Ranvier
   (c) Motor end-plate
   (d) Axon hillock
   (e) Neuromuscular junction

6. The neuron whose cell membrane receives neurotransmitter molecules released from the vesicles of another neuron:
   (a) Presynaptic
   (b) Beta cell
   (c) Acinar cell
   (d) Postsynaptic
   (e) Fascicle

7. Neuron A fires an action potential, exciting Neuron B to fire an action potential, which in turn excites Neuron C to fire an action potential. However, Neuron D is influenced by Neuron C, so that it does *not* fire an action potential. Neuron C is probably influencing Neuron D in this way:
   (a) Summation of twitches
   (b) Destruction of its neurotransmitters
   (c) Releasing GABA or glycine
   (d) Depolarizing its plasma membrane
   (e) Cutting away its myelin

8. The axon hillock's special significance:
   (a) Place where synaptic vesicles are stored
   (b) Contains the nucleus and most other organelles
   (c) Usual site of contact between neurons
   (d) Firing zone or trigger zone for a neuron action potential
   (e) Place where Baby Heinie exerts his demonic influence!

9. A neuron cell membrane is described as "resting polarized" when:
   (a) It is actively conducting a nerve impulse
   (b) No ions are crossing through it
   (c) The intracellular fluid is a negatively charged pole, while the ECF is a positively charged pole
   (d) The cell is totally incapable of further excitation
   (e) The ICF is a positive pole, while the ECF is a negative pole

10. An action potential is best visualized as:
    (a) A traveling wave of depolarization
    (b) A stationary tree that keeps growing and growing
    (c) Some sick giraffe without its spots!
    (d) A swarm of charged cations, continually flowing out of the neuron
    (e) An arrow flying straight toward its target

11. Myelin has an important role to play in:
    (a) Pumping poisonous materials out of nerve cells
    (b) Insulating neuron axons from excessive loss of electrical charge
    (c) Promoting homeostasis of blood Ca^{++} ion concentration
    (d) Delivering packets of neurotransmitters
    (e) Serving an enzyme-like role in speeding up cellular reactions

12. Nodes of Ranvier:
    (a) Represent bead-shaped chunks of the myelin sheath
    (b) Permit wave propagation via local circuit currents of ion flow
    (c) Are closely modeled by bare copper wires carrying electrons
    (d) Make up most of the white matter seen in gross anatomy
    (e) Allow impulse movement via saltatory conduction

13. The medial plantar nerve would most likely be stimulated by:
    (a) A feather stroking the middle of the sole of the foot
    (b) Some cold fish wrapped around your lips!
    (c) Hot beer being poured down your gullet!
    (d) A sharp nail being jabbed into the calf
    (e) Loud sound waves from a beating drum

14. A withdrawal reflex arc involving movements of the foot:
    (a) Involves only the ventral root of a spinal nerve
    (b) Waves from sensory receptors in the sole that descend to the spinal cord
    (c) A curved pathway of stimulation and automatic response
    (d) Voluntary lifting of the foot off some harmful or annoying stimulus
    (e) Activation of the brachial nerve, such that the impulses travel in an arc

15. The dorsal root ganglion represents:
    (a) White matter of the spinal cord
    (b) Cell bodies of sensory neurons lying just outside the spinal cord
    (c) Dendrites of motor neurons traveling up through the brainstem
    (d) Groups of sensory neuron cell bodies in the horns of spinal gray matter
    (e) Myelinated axons entering and leaving the body of the spinal cord

16. An interneuron is best described as:
    (a) A small, short association neuron between two larger neurons
    (b) One lower motor neuron that carries movement-related information
    (c) Some sensory neuron whose body lies in a horn of spinal gray matter
    (d) About the same thing as an axon hillock
    (e) Missing the usual supply of synaptic vesicles within its axon terminals

17. The spinothalamic tract has as its main function:
    (a) Transmission of ions from one spinal nerve to another
    (b) Carrying of nerve impulses from higher levels of the cord down to lower ones
    (c) Serving as an ascending pathway for many general body sensations on their way up to the cerebral cortex
    (d) Decussation of descending nerve fibers through the pyramids of the medulla oblongata
    (e) Providing anesthesia to the brain for otherwise intolerably painful sensory experiences!

18. The "heads" of the female clitoris and male penis:
    (a) Secreting "acorns" that supply vital reproductive hormones
    (b) Exocrine glands associated with production of germ cells
    (c) Abnormal, tumor-like swellings that interfere with sexual climax

    (d) Provide special motor pathways that relay in the thalamus

    (e) Glans that are named for their anatomic resemblance to real acorns

19. A common alternative or trade name for the hormone epinephrine:
    (a) Norepinephrine
    (b) ACh
    (c) Synapsin
    (d) ACTH
    (e) Adrenaline

20. The odd analogy made in the text to the human bloodstream as an "unstoppered sink" was helping to illustrate the fact that:
    (a) Once a hormone is secreted into the bloodstream, it never leaves
    (b) The secretion of hormones from glands is an on-and-off event
    (c) Releasing hormones from the hypothalamus stimulate the hypophysis
    (d) Some body cells are non-targets for particular hormones
    (e) Hormone molecules in the bloodstream are being continuously inactivated by the liver and then excreted into the urine by the kidneys

21. Antidiuretic hormone (ADH) owes its long name to its effect of:
    (a) Promoting secretion of H^+ ions into the urine
    (b) Enhancing reabsorption of glucose out of the kidney tubules
    (c) Protecting "Against Diarrhea" that is very "Heavy"
    (d) Control of blood glucose concentration
    (e) Increasing reabsorption of H_2O from the kidney tubules

22. Insulin is essential in human physiology for its job of:
    (a) Unloading oxygen into the red blood cells
    (b) Boosting the rate of facilitated diffusion of glucose out of the bloodstream, and into adipocytes, hepatocytes, and muscle fibers
    (c) Serving as a "Second Messenger" for anterior pituitary hormones
    (d) Increasing blood glucose concentration, by helping or "facilitating" its absorption from the stomach
    (e) Acting synergistically with cortisol, so as to provide the body with a high load of available glucose circulating through the bloodstream

23. The trophic hormones are all known for their effect of:
    (a) Stimulating the adrenal medulla to secrete testosterone
    (b) Promoting storage of glucose within liver cells as glycogen

(c) "Nourishing" or stimulating other endocrine glands

(d) Inhibiting or delaying the onset of the "fight-or-flight" response

(e) Relieving symptoms of starvation and dehydration

24. "Second Messengers" are often necessary, because:
    (a) Many hormone molecules are too large to enter their target cells
    (b) There is no real "First Messenger" available
    (c) Sometimes a glandular message is only imaginary, not real
    (d) Certain glands are starved for their normal supply of oxygen
    (e) The hypothalamus of the brain does not work without them

25. Releasing hormones are all characterized by their:
    (a) Common secretion from the epithelial cells of the anterior pituitary
    (b) Temporary storage, then "release" by the posterior pituitary
    (c) Tendency to split into several smaller hormones of similar function
    (d) Ability to stimulate the anterior pituitary to release trophic hormones
    (e) Property of exerting negative feedback control upon the rate of secretion of all hormones within the human organism

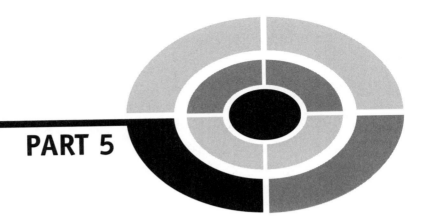

PART 5

Air and Fluid Transport: The Circulatory, Lymph, and Respiratory Systems

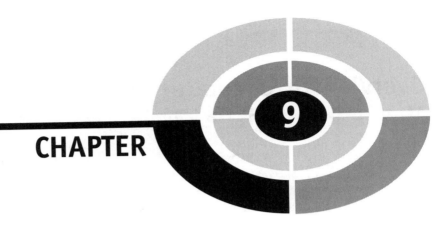

CHAPTER 9

Physiology of the Circulatory System: Mover of the Blood in a "Little Circle"

Back in Chapter 8, we talked of the importance of the blood circulation in distributing the hormones secreted by the various endocrine glands. Now it is time for us to focus upon the circulation and pumping of the blood itself.

Overview of Circulatory Physiology: Everything Moves in a "Little Circle"

Figure 9.1 (A) provides a general overview of the *circulatory* (**SIR**-kyuh-luh-**tor**-ee) or *cardiovascular* (**car**-dee-oh-**VAS**-kyuh-lar) *system*. In this organ system, the "heart" (*cardi*) pumps the blood in a "little circle" (*circul*), which travels through "little vessels" (*vascul*).

Figure 9.1 (B) shows the most important vessels within the cardiovascular (circulatory) system. Our close partner, *ANATOMY DEMYSTIFIED*, characterizes these different types of blood vessels in much more detail. Our primary job within the pages of *PHYSIOLOGY DEMYSTIFIED* is to try to explain how all of these vessels and the heart *work* together.

The Heart: A Double-Pump for Two Circulations

The human heart is a four-chambered double-pump. Figure 9.2 provides a brief look at its internal anatomy.

A BRIEF PEEK AT THE PULMONARY (RIGHT-HEART) CIRCULATION

The two chambers on the *right*-hand side of the heart are the pump for the *pulmonary* (**PULL**-mah-**nair**-ee) or "lung" circulation. These two chambers are named the *right atrium* (**AY**-tree-um) and the *right ventricle* (**VEN**-trih-kl). The right atrium (abbreviated as *RA*) is a small "entrance room" (*atri*) "present" (*-um*) at the top of the heart. When the walls of the right atrium (RA) contract, they squeeze the blood down into the right ventricle (abbreviated as *RV*). The right ventricle is a "little belly"-like cavity near the bottom of the heart.

But before the blood gets down into the right ventricle, it has to push open a one-way valve. This is called the *right atrio-ventricular* (**ay**-tree-oh-ven-**TRIK**-yew-lar) or *right A-V valve*. The name reflects the fact that this is the valve between the right "atrium" (*atrio*) and the right "ventricle" (*ventricul*).

Once inside the right ventricle, the blood is powerfully pumped up through the *right semilunar* (**sem**-ee-**LOO**-nar) *valve* or *right S-L valve*. The word semilunar "pertains to" (*-ar*) a "half" (*semi-*) "moon" (*lun*). This reveals that the right semilunar valve has flaps shaped like half-moons.

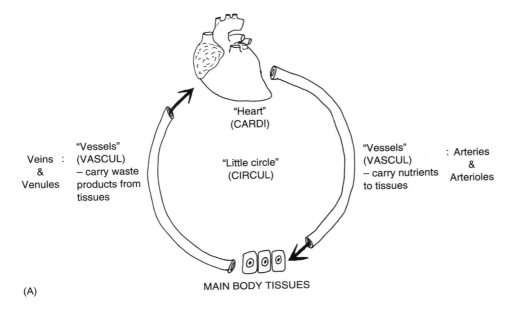

"Heart"
(CARDI)

Veins : "Vessels"   "Little circle"   "Vessels"   : Arteries
  &   (VASCUL)   (CIRCUL)   (VASCUL)   &
Venules   – carry waste   – carry nutrients   Arterioles
           products from   to tissues
           tissues

(A)   MAIN BODY TISSUES

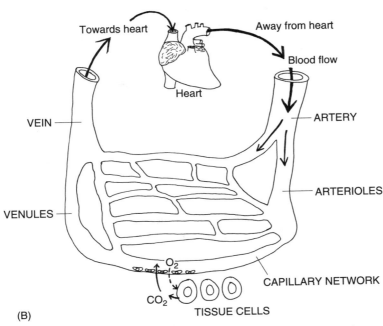

(B)

Fig. 9.1   An overview of the circulatory system and its parts. (A) A general overview of the
circulatory (cardiovascular) system. (B) The important types of vessels in the circulatory system.

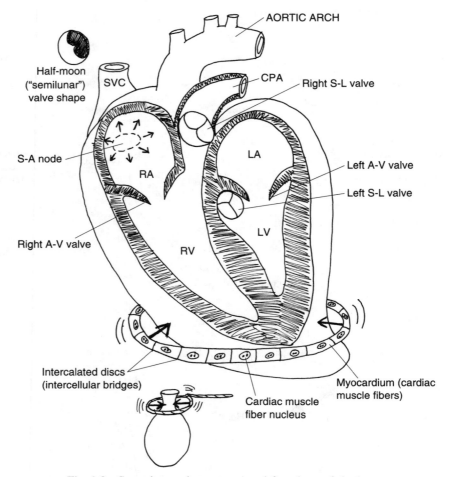

Fig. 9.2   Some internal structures and functions of the heart.

The right semilunar valve is located at the base of the *common pulmonary artery*, abbreviated as *CPA*. This artery is called a "common" one, because it serves as the trunk for the two smaller *right and left pulmonary arteries*, which carry the blood up into both lungs.

In summary, we have:

OUTPUT OF *PULMONARY*   RA→RIGHT A-V VALVE→RV→
(RIGHT HEART)   RIGHT S-L VALVE→CPA→
CIRCULATION: R&L PULMONARY ARTERIES→
                               R&L "LUNGS" (*PULMON*)

Organ System 1

A BRIEF PEEK AT THE SYSTEMIC (LEFT-HEART) CIRCULATION

The two chambers on the *left*-hand side of the heart are the pump for the systemic (sis-**TEM**-ik) or "*all organ systems*" (except for the lungs) *circulation*. (Now, look at the left side of the heart diagram pictured in Figure 9.2.) These two chambers are named the *left atrium* (abbreviated as *LA*) and the *left ventricle* (abbreviated as *LV*). When the walls of the left atrium (LA) contract, they push the blood down through the *left atrio-ventricular* or *left A-V valve*. Similar to what we saw for the right A-V valve, the left A-V valve is named for its position between the left *atrium* above it, and the left *ventricle* below it.

Once inside the left ventricle, the blood is powerfully pumped up through the *left semilunar* or *left S-L valve*. Again mirroring the situation on the right side of the heart, the flaps of the left semilunar valves also resemble half-moons.

After the blood passes through the open flaps of the left S-L valve, it enters the *aortic* (ay-**OR**-tik) *arch*. "Oh, is this like the Golden Arches, Professor? You know . . . the place where they sell all those yummy fries and hamburgers?" Well, Baby Heinie, in a sense you are correct! In this case, the aortic arch might be called the *Red* Arch, since it is filled with *red*, oxygen-rich blood.

"Where is all this bright red blood *going*, Professor?" Now, Baby Heinie, since it is a Red *Arch*, you *ought-er* know that the aortic arch is the first part of the *aorta* (ay-**OR**-tah)! And the blood in the aorta (and its aortic arch) is literally being "raised or lifted up" (*aort*) out of the left side of the heart.

The aorta, and its first curved part, the aortic arch, is the Great Mother Artery for the systemic circulation. There are dozens of smaller *systemic arteries* (such as the brachial artery and femoral artery) that supply blood to the organs of all the major body *systems* (except for the lungs). This systemic output includes, for example, blood flowing to our skeletal muscles, our bones, our skin, our brains, and our many other viscera (internal organs).

In summary, we have:

OUTPUT OF *SYSTEMIC* (LEFT HEART) CIRCULATION:	LA→LEFT A-V VALVE→ LV→LEFT S-L VALVE→ AORTIC ARCH (AORTA)→ THE SYSTEMIC ARTERIES→ ORGANS OF ALL BODY "SYSTEMS" (*SYSTEM*) EXCEPT FOR LUNGS

Organ System 2

Cardiac Pacemaker Tissue: "Spark Plugs" in the Heart Wall

Earlier chapters talked about the way in which skeletal muscle was stimulated to contract by means of motor neurons. Specifically, motor neurons have their action potentials (traveling waves of electrochemical or ionic excitation) *first*. After they release excitatory neurotransmitters from their axon terminals, the neurons stimulate the skeletal muscle fibers to fire their own action potentials, *second*. It is only after being stimulated by motor neuron "spark plugs", then, that skeletal muscle fibers are excited to contract.

"ONE FOR ALL, AND ALL FOR ONE": THE MYOCARDIUM AS THE THREE MUSKETEERS

Now, cardiac muscle fibers, like their other striated (cross-striped) relatives, the skeletal muscle fibers, have to be stimulated or excited before they contract. In the case of the cardiac muscle tissue, however, the source of excitation lies *within* the muscle tissue itself.

This very special source of excitation is called the *cardiac pacemaker tissue*. It is also called *nodal* (**NOHD**-al) *tissue*, because it looks like a little oval "knot" (*node*). The most important portion of the cardiac pacemaker (nodal) tissue is called the *sinoatrial* (**sigh-no-AY**-tree-al) or *S-A node*. The word sinoatrial indicates the anatomic location of this fairly oval, "knot"-shaped, area. (Take a look back at Figure 9.2.)

The S-A node lies in the outer wall of the right "atrium" (*atrial, A*). It is the area just below the entrance of the *superior vena* (**VEE**-nah) *cava* (**KAY**-vah) – a major *venous* (**VEE**-nus) *sinus* (*sino-, S*). Hence we have the phrase *sinoatrial* (*S-A*). (A sinus, in general, is a large vein that is shaped like a "hollow bay" [*sin*], and holding a considerable volume of blood.)

The sinoatrial node is known as the *primary cardiac pacemaker*. The cardiac muscle fibers within this node can, somewhat like a spark plug, spontaneously depolarize or excite themselves. Thus the atria (upper entrance rooms) on either side of the heart are the first chambers to contract, because they are excited by the *cardiac pacemaker cells* in the S-A node. The cell membranes of these very special muscle fibers are able to become spontaneously leaky to sodium (Na^+) and other cations, at a certain rate or rhythm. This happens because the proteins in the membranes of the pacemaker cells tend to shift around, allowing positively charged ions to diffuse inward.

Hence, the pacemaker cells excite or "turn themselves on" at a certain rate or rhythm. They fire off a series of *cardiac muscle action potentials*. The speed with which these pacemaker cells spontaneously excite or depolarize themselves, and fire off cardiac muscle action potentials, in turn sets the stage for the *resting heart rate (HR) or pulse rate*.

There are no synapses between adjacent cardiac muscle fibers. Figure 9.2 demonstrates that the cardiac muscle fibers are located within the *myocardium* (**my**-oh-**KAR**-dee-um). The myocardium is literally the "muscle" (*myo-*) tissue "present" (*-um*) within the "heart" (*cardi*) wall.

All of the cardiac muscle fibers in the myocardium are tightly wrapped around the heart and its chambers, in a series of whorls or spiral patterns. The myocardium can largely be thought of as a *functional syncytium* (sin-**SIT**-ee-um). The word syncytium exactly translates from Latin to mean "a condition of" (*-um*) "cells" (*cyt*) "together" (*syn-*). By a functional syncytium, we mean that the adjacent cardiac muscle fibers within the myocardium tend to *act together*, at about the same time, almost as if all of them were a single muscle fiber.

"Oh, so it's kind of like the theme of the Three Musketeers – 'One for all, and all for one!'" That is correct, Baby Heinie. "But if the cardiac muscle fibers are really *separate* cells, rather than a *single* cell, how they are able to contract so closely together?"

The reason is that the adjacent cardiac muscle fibers in the heart wall are linked by *intercalated* (**in**-ter-**KAY**-lay-ted) *discs*. Back in Figure 9.2, we see these intercalated discs "inserted between" (*intercalat*) neighboring cardiac muscle fibers. These intercalated discs appear as thick dark lines (considerably thicker than the regular cross-stripes or striations) within cardiac muscle fibers, when they are viewed through a compound microscope. Their main function is to act as *intercellular* (**in**-ter-**SELL**-you-lar) *bridges* – bridges "between" the cardiac muscle "cells." Therefore, when the S-A node fires off a cardiac muscle action potential, this traveling wave of excitation moves from one cardiac muscle fiber to another, across the intercalated discs connecting them end to end. Eventually, both atria become excited and contract. And finally, both ventricles become excited, then contract.

Major Events of the Cardiac Cycle

After the chambers of the heart are excited, they go into a state of *systole* (**SIS**-toh-**lee**) or "contraction" and emptying. It is during systole that the blood is pumped from each of the ventricles and out into their major arteries.

After systole comes *diastole* (die-**AH**-stoh-**lee**). Whereas systole is the contracting and emptying phase of each chamber of the heart, diastole is the "relaxation" and filling phase of each chamber.

AN OUTLINE OF THE CARDIAC CYCLE

The *Cardiac Cycle* is one heartbeat, or one complete sequence of contraction (systole) plus relaxation (diastole) of all four chambers of the heart. It is convenient for us to construct two circles to represent the Cardiac Cycle – a smaller one tucked inside a larger one (Figure 9.3). The smaller, inner circle represents the activity of both the right and left atria (upper heart chambers) during a typical Cardiac Cycle. And the larger, outer circle models the activity of the right and left ventricles (lower heart chambers) during a typical Cardiac Cycle.

You can measure the number of Cardiac Cycles (heartbeats) you are having in your own body at the present time, by palpating (touching with your

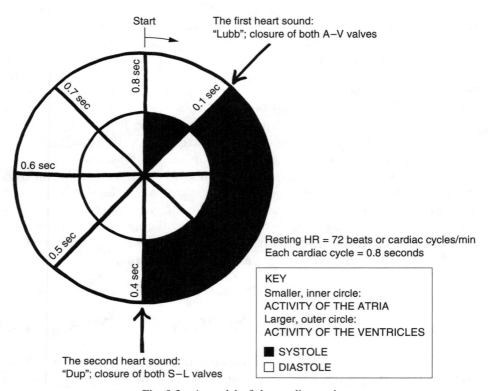

Fig. 9.3 A model of the cardiac cycle.

fingertips) the *radial artery* in the wrist. [**Study suggestion:** Using several fingers of one hand, gently palpate for the radial artery in the wrist on the other side of your body. It is located just proximal to the base of the thumb, within a shallow, lengthwise indentation. You should be able to feel a steady thumping of the outer wall of the radial artery against your fingertips. Can you find it? Each thump you feel is a surge of blood passing through the radial artery during a single Cardiac Cycle. Now, count the number of "thumps" you feel against your fingertips, for a full 60 seconds. The result is a very important body parameter – heart rate (HR) or pulse rate (PR) measured in number of beats or Cardiac Cycles/minute. Is this an anatomical parameter or a physiological one? Why?]

Suppose that you counted thumps and found you had a resting HR of 72 beats or Cardiac Cycles/minute. This is about what the average adult has for a resting heart rate. Dividing 60 by 72, we calculate that each Cardiac Cycle lasts about 0.8 seconds. Applying this to Figure 9.3, we will cut our two circles into halves, into quarters, and finally into eighths. Each of the eight slices of our two Cardiac Cycle "pies" (big ventricular circle, little atrial circle), then, lasts for 0.1 second. Let us dig into Figure 9.3, and see what happens inside the heart, now, during each of these 8 "pie slices"!

During the first 0.1 second, both atria go into systole. Note that during this same period, both ventricles are in diastole. Both A-V valves are open. Therefore, blood is pumped from the atria, down through the A-V valves, and into both resting ventricles.

A LONG "LUBB": THE FIRST HEART SOUND MEANS BOTH A-V VALVES ARE CLOSING

A *stethoscope* (**STETH**-oh-skohp) is "an instrument used to examine" (-*scope*) the "chest" (*steth*) for various sounds. When a stethoscope bell is placed against the chest, a health professional may perform *auscultation* (**aws**-kul-**TAY**-shun) – the "process of listening" (*auscult*).

After the first 0.1 second, auscultation allows us to hear the *first heart sound, "lubb."* The "lubb" sound is a rather dull, *prolonged* vibration (so the "lubb" has *two b's* in it, rather than only one). "Lubb" is created by the closure of both A-V valves at the beginning of *ventricular* (ven-**TRIK**-you-lar) *systole*. When the ventricles contract, they push blood up beneath the flaps of the A-V valves, snapping them shut from below. The benefit, of course, is that there is no leakage or *regurgitation* (re-**gur**-jih-**TAY**-shun) of blood from the ventricles back up into the atria.

Instead, as the ventricles powerfully contract, they push both semilunar (S-L) valves open. Blood is thus pushed up from the right ventricle, through the right S-L valve, and into the common pulmonary artery. On the other side of the heart, blood is pushed up from the left ventricle, through the left S-L valve, and into the aortic arch. This period of ventricular systole lasts for a total of 0.3 seconds. During this time, blood is being pushed up out of the ventricles and into the other vessels and organs supplied by the pulmonary and systemic circulations.

A SHORT "DUP": THE SECOND HEART SOUND MEANS BOTH S-L VALVES ARE CLOSING

Ventricular systole provides a vital, life-giving stream of blood to the body's organ systems. But, as the old saying goes, "All good things must come to an end." Specifically, the end of ventricular systole comes at 0.4 seconds.

At this time, the *second heart sound* can be heard using auscultation. It is a *short*, sharp "*dup*" sound. (Since "dup" is *short*, it ends with only *one p*.) The second heart sound ("dup") reflects the closing of both semilunar valves at the beginning of ventricular diastole. The blood in the common pulmonary artery/ aortic arch above each ventricle starts to fall back downward, due to the force of gravity. As the blood falls down, it catches the edges of the semilunar valve flaps and slams them shut, thereby preventing a back-leak or regurgitation of blood into the ventricles. [**Study suggestion:** Glance back at Figure 9.2 to help you visualize what is happening during the two heart sounds.]

In summary of the two heart sounds:

Organ 1

THE FIRST HEART   =   Sound made by closure of both *A-V valves*
    SOUND ("*lubb*")   at the beginning of ventricular *systole*

THE SECOND HEART   =   Sound made by closure of both *S-L valves*
    SOUND ("*dup*")   at the beginning of ventricular *diastole*

THE LAST HALF OF THE CARDIAC CYCLE: EVERYONE'S TAKING A WELL-DESERVED NAP!

Please observe that, after the first 0.4 seconds, all four chambers of the heart are in a state of diastole. Neither the ventricles nor the atria are contracting. Therefore, for fully the last half of each Cardiac Cycle, all chambers of the heart are resting!

The EKG: Patterns of Ion Flow Associated with the Cardiac Cycle

The technique of auscultation with a stethoscope can allow us to *hear* certain abnormalities (such as heart murmurs) occurring within the Cardiac Cycle. But another technique – the *electrocardiogram* (ee-**lek**-troh-**KAR**-dee-oh-**gram**) – can allow us to *see* other abnormalities for ourselves.

The word electrocardiogram is abbreviated as either *ECG* or *EKG*. The letter *K* comes from *kardi*, which is Greek for "heart." (We will use *EKG* as our abbreviation, because we think it sounds better!) An electrocardiogram (EKG) is literally a "graphical record" (-*gram*) of the "electrical activity" (*electro*) of the "heart" (*cardio*). This electrical activity shows up as tracings, called an *EKG wave pattern*, either on moving graph paper, or perhaps upon a computer monitor screen.

The basic EKG recording set-up is illustrated in Figure 9.4(A). At least three *recording electrodes* are placed upon the skin of a patient – usually on the right wrist, left wrist, and left ankle. (Several chest electrodes are often added.) These electrodes are then attached by electrical wires to an *electrocardiograph* (ee-**lek**-troh-**KAR**-dee-oh-**graf**), or *EKG machine*. A recording electrode attached by a wire to the EKG machine is technically called an *EKG lead*.

WHAT DO EKG WAVES REALLY REPRESENT?

Figure 9.4(B) shows a sample EKG wave pattern recorded from *Lead I*. Notice that there is a dark, horizontal *baseline*. The EKG waves, then, are seen as either upward or downward deflections or bumps, from this horizontal baseline. "Professor Joe, are these EKG waves really just cardiac muscle *action potentials*, which we have also called traveling waves?" No, they are *not* actually cardiac muscle action potentials, because those action potentials travel from the S-A node pacemaker, through the heart muscle itself. The electrodes we are using for recording the EKG are *not* even *on* the heart surface, are they? "No, they're on the *skin* surface."

That is correct. Now, we know that action potentials are really traveling waves of ions – atoms with a net electrical charge – such as sodium (Na^+) ions. The human heart is basically surrounded by the ECF (extracellular fluid). The ECF is essentially a *saline* (**SAY**-leen) or "salty" solution of water and 0.9% NaCl solute. This sodium chloride in the ECF breaks

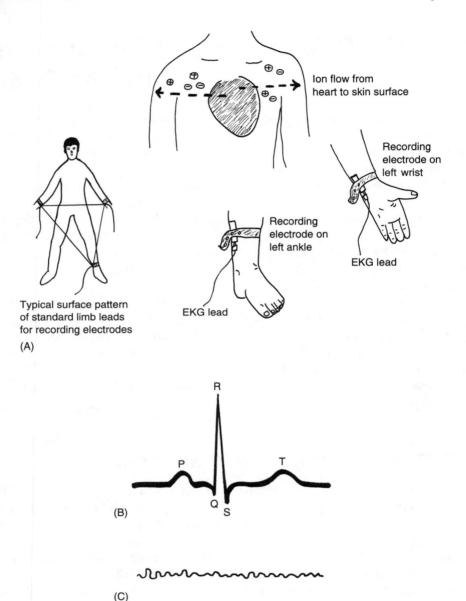

Fig. 9.4 An EKG recording set-up and associated wave patterns. (A) The basic EKG standard limb lead arrangement that picks up ion flow from the heart. (B) A normal EKG wave pattern associated with a single cardiac cycle (heartbeat). (C) A highly disordered EKG pattern reflecting the electrical chaos of ventricular fibrillation.

down into millions of charged Na$^+$ and Cl$^-$ ions. Together with various other cations and anions dissolved in water, they compose the "internal sea" or internal environment, which lies deep to the surface of the skin.

Whenever the S-A node becomes depolarized (electrochemically or ionically "excited"), it fires off a series of cardiac muscle action potentials. These action potentials quickly move throughout the heart (a functional syncytium), because the intercalated discs allow the action potentials to travel from one cardiac muscle fiber to another. After they are excited or depolarized by the action potential, the cardiac muscle fibers contract. This contracting movement makes the heart chambers pump the blood.

While all this conducting of cardiac muscle action potentials is going on *within* the heart wall, a large group of ions are diffusing *away from* the surface of the heart, and through the ECF. Remember (Chapter 4) the general principle that "like dissolves like." Therefore, the charged ions diffusing away from the surface of the heart, readily mix and pass through the charged, saline, watery ECF around the heart.

Eventually, the charged currents of ions diffusing from the heart surface, reach the skin surface. Here they are detected by the recording electrodes of the EKG machine.

Therefore, the EKG wave patterns actually represent the **pattern of ions** *flowing* **from** *the surface of the* **heart**, *and* **onto** *the surface of the* **skin.** The EKG waves *indirectly* represent the electrochemical activity of the heart, because the shape, size, and other characteristics of the waves are strongly influenced by the ionic events going on within the heart. [**Study suggestion:** Picture this abstract situation using a helpful model. An old-fashioned movie projector is sitting in the back of a classroom. The movie projector is the heart. The electrochemical or ionic events occurring in the heart, then, are a movie on film. The heart (movie projector) casts a visual image through the air (ECF) and onto a screen. The screen is the skin surface. The visual images being viewed on the screen (observed wave patterns of an EKG) are not the true events actually occurring within the movie projector (heart). The true events are a film moving through a series of gears and sprockets. But the visual pattern being projected onto the screen of the skin – the waves of the EKG – very much depend on what is happening, electrochemically or ionically speaking, within the walls of the heart.]

Organ 2

A NORMAL EKG WAVE PATTERN

Because the electrodes in the EKG leads are arranged in a consistent manner, a consistent series of waves usually occurs within the EKG record (see Figure

9.4, B.) These waves are identified in alphabetical sequence, starting with the letter *P*. (The selection of the letters from *P* onward is pretty arbitrary, and does not, itself, mean anything.) The first wave seen is usually the *P wave*. The P wave is a small upward bump that indirectly represents the depolarization (excitation with cardiac muscle action potentials) of both the right and left atria. (The two atria are depolarized first, because the S-A node is found in the right atrium. These two upper chambers are the heart parts located closest to the S-A node pacemaker.)

Next in sequence is the *Q-R-S wave complex*. This combination of three waves represents the depolarization (excitation) of both ventricles. Just as the P wave comes before the systole (contraction) of the atria, then, the Q-R-S wave complex comes before the systole of the ventricles and their pumping out of blood into major arteries.

The next common event is the *T wave*. This wave represents the *repolarization* (**REE-poh**-lar-ih-**zay**-shun) of the ventricles. By this it is meant that the ventricles are going back to being resting "polarized," once "again" (*re-*). This was the state they were in before they were *depolarized* – "taken away from" (*de-*) a resting "polarized," non-excited state, by cardiac action potentials from the nodal tissue.

Taken altogether, the resulting P, QRS, T combination represents the complete EKG wave pattern usually associated with each beat, Cardiac Cycle, or "lubb dup" of the heart.

Organ 3

In summary:

NORMAL EKG WAVE PATTERN (Pattern of ion flow associated with one beat, Cardiac Cycle, or "lubb dup" of the heart)	=	P WAVE (*Depolarization* or excitation of both *atria*)	+	QRS WAVE COMPLEX (*Depolarization* or excitation of both *ventricles*)	+	T WAVE (*Repolarization* or "*return*" of *ventricles* to resting *polarized* state)

VENTRICULAR FIBRILLATION

During ventricular fibrillation (**feye-bril-AY-shun**), however, the normal EKG wave pattern disappears and is replaced by a highly disordered state of random electrical chaos (Figure 9.4, C). The ventricles beat in a totally uncoordinated manner, so that the heart looks like a bag of writhing worms, or a bunch of independently contracting "little fibers" (fibrils).

Organ 1

The Work of the Heart: Stroke Volume and Cardiac Output

Once the heart is properly excited and contracting normally, what influence does this have upon the *rest* of the body? What we are really asking here is something about the *work* being done or achieved by the heart.

Three critically important physiological parameters associated with the work of the heart are the heart rate (HR), the *stroke volume* (*SV*), and the *cardiac output* (*CO*). We have already learned that the heart rate (also called pulse rate) represents how *fast* the heart is contracting. It is measured as so many beats or Cardiac Cycles per minute. Recall that the average heart rate for a resting adult is about 72 beats/minute.

STROKE VOLUME AND THE STRENGTH OF HEART CONTRACTION

The stroke volume, in contrast, indirectly indicates how *strongly* the heart muscle (myocardium) is contracting. The stroke volume (SV) is the *volume* of blood ejected from the ventricles with each beat or *stroke* of the heart. The SV is usually measured as so many milliliters (ml) of blood per "stroke" or beat of the heart. A typical value for SV in a resting adult is 70 ml/beat or stroke. [**Study suggestion:** Get a big red balloon to model a ventricle of the heart. Fill the balloon with water. Now, aim its stem over a sink. Using one hand, "stroke" or compress the balloon one time. The volume of water you see being ejected from the balloon stem – the major connected artery – represents the stroke volume. Now, stroke the balloon (heart ventricle) one more time, but with greater strength or force. What do you observe about the volume of fluid (the stroke volume) being squirted out into the sink?]

In general, the stroke volume of the left ventricle is about equal to the stroke volume of the right ventricle. This means that, at rest, about 70 ml of blood are being pushed out of the *left* ventricle and into the *aortic arch*. And at the same time, about 70 ml of blood are also being pushed out of the right ventricle and into the *common pulmonary artery*. Thus, both sides of the heart are usually contracting with about the same amount of strength or force.

To help us see this, let us visualize the pulmonary circulation as the *right heart or lung pump*, and the systemic circulation as the *left heart or systemic pump*. As shown in Figure 9.5, these two pumps are connected *in series*, one right after the other.

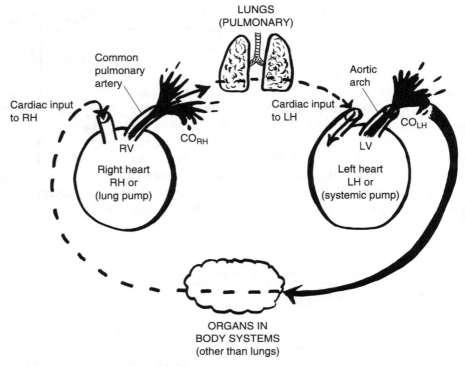

Fig. 9.5 "Balloon" models for observing the work of the heart.

CARDIAC OUTPUT AS THE PRODUCT OF SV AND HR

Since both sides (pumps) of the heart are connected in series, their cardiac outputs must be equal. Cardiac output (CO) represents the volume of blood being ejected from each pump or side of the heart, every minute. The cardiac *output* of the *right* heart eventually becomes the cardiac *input* of the *left* heart. Likewise, the cardiac *output* of the *left* heart eventually becomes the cardiac *input* of the *right* heart. This is true because of the fact that both heart pumps are connected in series, such that all of the fluid *coming out* of one pump, eventually *goes into* the other pump, and vice versa. [**Study suggestion:** Go back and get your big red balloon. Fill the balloon with water, and let it represent the right heart (actually, the right ventricle). Now, get *another* big red balloon, and fill it with water. This second balloon, of course, models the left heart (actually, the left ventricle). Pretend that the two balloons (sides of the heart) are connected together. Keep stroking the "right heart" again and again for one minute, and you have the CO of the "right heart." At the same time, keep stroking the "left heart" again and again for one minute, and you have the CO of the "left heart."]

The CO of the right heart, therefore, must approximately equal the CO of the left heart, or else one side of the heart will "fall behind" the work being accomplished by the other side of the heart!

We have the following general equation that expresses the relationship among HR, SV, and CO:

CARDIAC OUTPUT = STROKE VOLUME ×   HEART RATE
  (CO) in liters or   (SV) in ml/stroke or   (HR) in no. of beats or
      ml/min   beat of heart   strokes/minute

For practice in computing cardiac output (CO), let us multiply a typical resting adult stroke volume (SV) of 70 ml/beat, times the resting adult heart rate (HR) of 72 beats/minute. [**Study suggestion:** Carry out this multiplication on your own, and then check your answer with the one given in the following sentence.] The answer is CO = 70 × 72 = 5,040 ml/minute. For convenience, divide this product by 1,000, and get CO = 5.04 liters/minute.

Organ 4

THE PRACTICAL SIGNIFICANCE OF CARDIAC OUTPUT

Cardiac output is a critically important body parameter, because it represents the total amount of *tissue perfusion* (per-**FYOO**-zhun) – the "pouring" of blood "through" the tissues – that occurs every minute of our life! Thus, it represents the total amount of blood that is bathing our tissue cells every minute, providing them with their necessary lifeline of oxygen, glucose, and other vital nutrients. Likewise, it serves as the amount of blood that carries poisonous waste products away from the tissue cells, which result from their active metabolism.

To summarize the importance of CO and its two main determinants, SV and HR, we state that:

CO	=	SV	×	HR
(The total amount of		(The amount of		(The number of
blood perfusing the body		blood perfusing the		times that the
tissues every minute of		tissues with every		heart beats every
life)		heartbeat)		minute)

Organ 5

The Importance of Controlling Cardiac Output

Since cardiac output represents the total amount of blood perfusing or bathing our tissue cells every moment of our life, it would be interesting to find out some normal values for this critical physiological parameter. As Figure

9.6 (A) shows, the normal range for CO is from about 4 to 8 liters/minute. The average or set-point value is about 5 liters/minute. The value of 5 liters is about the total amount of blood volume contained within the vessels of an average-sized adult. Therefore, at rest, just about the entire blood volume is pumped throughout the body as the cardiac output, every single minute!

If homeostasis of CO is maintained, then the result is *tissue normoperfusion* (**nor**-moh-per-**FYOO**-zhun) – a "normal" amount of blood "pouring through" or bathing our body tissues.

Under conditions of exercise or stress, both HR and SV tend to increase. Thus, CO will also tend to increase toward its upper normal limit of about 8 liters/minute, because it is the product of a higher HR multiplied by a higher SV.

CARDIAC ARREST AND TISSUE ISCHEMIA (HYPOPERFUSION)

"What happens if the person has a heart attack, Professor Joe?" Well, in a cardiac arrest, the heart essentially slows down or even stops pumping and beating. Specifically, both HR and SV greatly decline. This will cause the resulting CO to fall far below its lower normal limit of 4.0 liters/minute (see Figure 9.6, B).

The result is *tissue hypoperfusion* (**high**-poh-per-**FYOO**-zhun), which is more commonly described as *tissue ischemia* (is-**KEE**-mee-ah). Hypoperfusion is a "below-normal or deficient" amount of blood "pouring through" the tissues. A synonym (word with similar meaning) is ischemia, which comes from Greek for "a holding back of blood."

If tissue hypoperfusion (or tissue ischemia) is very prolonged or severe, then *tissue necrosis* (neh-**KROH**-sis) or an "abnormal condition of" (-*osis*) tissue "death" (*necr*) may result. Thus, cardiac output must be maintained within its normal range, if adequate tissue perfusion and circulation of blood throughout all parts of the body are to be achieved.

Blood Pressure and Blood Flow

We have discussed such important physiological parameters as cardiac output, heart rate, and stroke volume. But at least two vital physiological parameters associated with the cardiovascular system still need to be examined. These are *blood pressure* (*BP*) and *blood flow*.

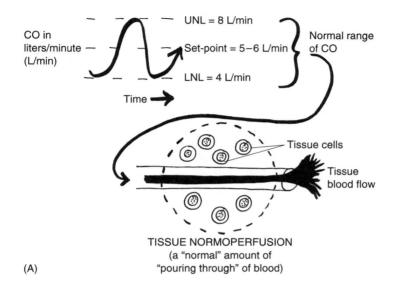

(A)

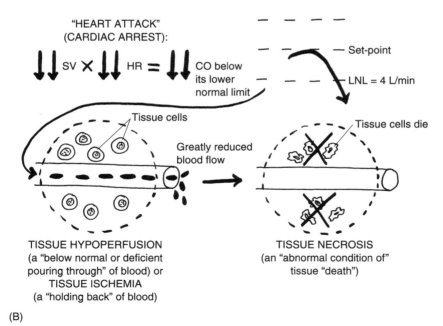

(B)

Fig. 9.6 Cardiac output: just enough, or too little? (A) CO in its normal range results in tissue normoperfusion. (B) CO below its normal lower limit results in tissue hypoperfusion or ischemia.

In general, blood pressure is a pushing force exerted against the blood and against the walls of the blood vessels (see Figure 9.7). The BP is at its highest in the major arteries attached directly to the heart (such as the aortic arch and common pulmonary artery). The BP then progressively decreases with greater distance from the heart.

Blood pressure is hugely significant because it is the main force causing blood flow. There is a *blood pressure gradient* (**GRAY**-dee-unt), a series of downward "steps" from a region of higher to a region of lower blood pressure. Within the systemic circulation, for example, the BP is highest within the aortic arch. As the aortic arch progressively branches into smaller and smaller arteries farther away from the heart, the BP also declines. This creates a blood pressure gradient or series of downward steps in blood pressure in vessels farther from the heart. Hence, the blood is pushed *down* a BP gradient as it leaves the heart, so that it flows into lower-pressure arteries in a series of steps.

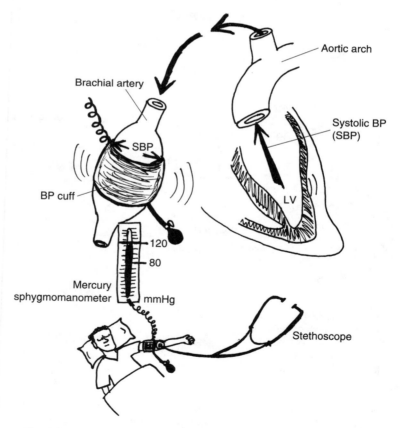

Fig. 9.7 Blood pressure (BP), blood flow, and the brachial artery.

One of these farther arteries, the brachial artery (you may remember), is in the upper "arm" (*brachi*). The brachial artery is the most frequent site used for taking someone's blood pressure, utilizing a stethoscope and *sphygmo-manometer* (**sfig**-moh-muh-**NAHM**-uh-ter). A sphygmomanometer is literally "an instrument used to measure" (*-meter*) the "throbbing pulse" (*sphygmo*) at certain "intervals" (*mano*).

To be sure, one does hear a dull throbbing sensation when the bell (expanded end) of a stethoscope is placed over the brachial artery and the arm cuff of a sphygmomanometer device is inflated around it. The sphygmo-manometer is usually marked off in units of *millimeters of mercury*, abbre-viated as *mmHg*. (Hg is the chemical symbol for the element mercury.)

As the air is slowly let out of the inflated arm cuff, the first dull throbbing noise one hears through the stethoscope is called the *systolic* (**sis-TAHL**-ik) *blood pressure*, or *SBP*. The systolic blood pressure (SBP) is the pressure created by the systole (contracting and emptying phase) of the left ventricle of the heart. A slug of blood (the stroke volume) is pushed out of the left ventricle with its contraction. Flowing progressively down a BP gradient, this slug of blood finally enters the brachial artery. The blood bulges out the artery somewhat as it passes through it, thereby creating a thumping sound. For a resting adult, the systolic blood pressure (SBP) is usually recorded as about 120 mmHg.

The *diastolic* (**DIE**-ah-stahl-ik) *blood pressure*, or *DBP*, is the blood pres-sure associated with the diastole (relaxing and filling) phase of the left ven-tricle. Upon thoughtful reflection, you might well ask, "Why isn't the diastolic BP just 0 mmHg? Isn't diastole the resting and filling phase, when the ventricle isn't even contracting or creating any blood pressure? So, why should there even be any diastolic BP at all?"

Good question! Glance back at Figure 9.7. Note that the brachial artery bulges out with the force of the systolic BP against its walls. Now, when the ventricle stops contracting, and diastole begins, there is a powerful elastic recoil, or snapping-back force, created by the stretched brachial artery wall coming back to its non-stretched shape. This force of elastic recoil (from the powerful snapping back of the stretched brachial artery wall) is what creates the diastolic BP. The diastolic blood pressure is then, in a sense, a residual or left-over blood pressure. It is the force of the blood pressure temporarily "stored" in the bulged-out walls of the brachial artery during systole. Because it is a residual BP, the diastolic BP is normally significantly lower than the systolic BP. It averages about 80 mmHg in a resting adult.

The stethoscope and sphygmomanometer measure a resting adult's blood pressure as about 120/80 mmHg. (You can think of this as the set-point, or long-term average value, for the blood pressure parameter.) The SBP is 120, the DBP is 80. The diastolic BP is usually recorded as the level of blood pressure where the dull thumping sound heard through the stethoscope just disappears. [**Study suggestion:** Ask yourself, "Why does the thumping sound disappear just below the recorded level of the diastolic BP?" *Hint:* Think about what is happening to the wall of the brachial artery.]

Hypertension Versus Hypotension

We have been talking about the normal blood pressure, which has an average or set-point level of about 120/80 mmHg. The normal BP can rise to an upper normal limit of approximately 140/90 mmHg, and it can fall down to a lower normal limit of about 100/60 mmHg. The distance between these upper and lower normal limits, of course, creates the *normal range for blood pressure*. A staying of the BP within its normal range is technically called *normotension* (**NOR**-moh-**TEN**-shun). We use this term because blood "pressure" represents the amount of *tension* exerted against the blood. Figure 9.8 (A) illustrates a state of normotension, that is, a state of relative constancy or homeostasis of blood pressure within its normal range over time.

Unfortunately, however, BP does not always remain within its normal range. Say that a lumberjack accidentally cuts his brachial artery with a chainsaw while felling a large oak tree. The blood spurts out of the artery in hot red jets, resulting in a severe and possibly fatal *hemorrhage* (**HEM**-uh-rij) – a "bursting out" (*-orrhage*) of "blood" (*hem*). When so much blood is lost, there isn't much blood left to press against the arterial wall. Consequently, blood pressure steeply declines. It may even reach a state of *hypotension* (**HIGH**-poh-**TEN**-shun). Hypotension is a condition of "below-normal or deficient" (*hypo-*) blood "pressure" (*tens*). Specifically, hypotension is a blood pressure significantly less than 100/60 mmHg (Figure 9.8, B).

At the opposite extreme is *hypertension* (**HIGH**-per-**TEN**-shun) – an "excessive or above normal" (*hyper-*) blood pressure. Hypertension is a blood pressure significantly above the upper normal limit of approximately 140/90 mmHg (Figure 9.8, C).

"So what if a person has hypotension or hypertension?", a skeptic might inquire. With hypotension, the person may easily faint. And if the condition is severe, there may be *circulatory shock* or *coma*, due to a lack of blood pressure pushing blood up to feed the brain. In the case of hypertension, the chronically above-normal pressure may overstretch and thin out the walls of arteries. This thinning and ballooning out creates *aneurysms* (**AN**-yuh-**riz**-

Tissue 1

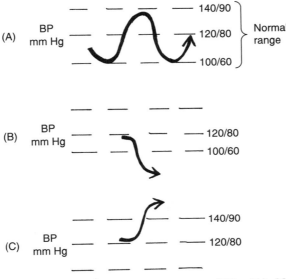

Fig. 9.8 The three possible states of blood pressure (BP). (A) Normotension. (B) Hypotension. (C) Hypertension.

ums). Aneurysms are abnormally "widened up" arteries, which are highly prone to being ruptured. And when an aneurysm ruptures, there may be a large amount of internal bleeding. Persons suffering a stroke, or *cerebrovascular* (seh-**REE**-broh-**VAS**-kyoo-lar) *accident*, for instance, may well have experienced a ruptured aneurysm of their cerebral blood vessels covering the brain. Whatever particular body functions the oxygen and blood-deprived brain area carried out are then partially or totally lost.

Quiz

Refer to the text in this chapter if necessary. A good score is at least 8 correct answers out of these 10 questions. The answers are listed in the back of the book.

1. The right-hand side of the heart is functionally described as the:
   (a) "Lubb dup" place
   (b) Systemic pump
   (c) Coronary circulation
   (d) Pulmonary pump

2. When the left ventricle contracts, it pushes blood up through the:
   (a) Left semilunar valve
   (b) Left atrio-ventricular valve
   (c) Right auricle
   (d) CPA

3. The primary cardiac pacemaker in the heart wall:
   (a) Myelin internode
   (b) P wave
   (c) Aortic arch
   (d) Sinoatrial node

4. _____ discs help make the cardiac muscle behave as a functional syncytium:
   (a) Striated
   (b) Smooth
   (c) Intercalated
   (d) Imaginary

5. The resting and filling phase of each chamber in the heart:
   (a) Systole
   (b) SV
   (c) Diastole
   (d) Contraction

6. Measured in number of beats/minute:
   (a) BP
   (b) HR
   (c) CO
   (d) MVP

7. The first heart sound represents the:
   (a) Squeezing noises of the atria
   (b) Ventricular repolarization
   (c) Closure of both A-V valves
   (d) Opening of both S-L valves

8. The EKG wave patterns essentially reflect:
   (a) Cardiac muscle action potentials occurring within the heart wall
   (b) Turbulent heart murmurs through damaged valves
   (c) Imaginary tracings having no real relationship to cardiac physiology
   (d) Patterns made by ions flowing from the heart to the skin surface

9. Indirectly indicates the strength of heart contraction:
    (a) P wave
    (b) Ventricular fibrillation
    (c) SV
    (d) Pulse rate

10. CO is calculated as:
    (a) TPR − SBP
    (b) DBP + SV
    (c) HR × BP
    (d) SV × HR

Body-Level Grids for Chapter 9

Several key body facts were tagged with numbered icons in the page margins of this chapter. Write a short summary of each of these key facts into a numbered cell or box within the appropriate *Body-Level Grid* that appears below.

Physiology and *Biological Order* **Fact Grids for Chapter 9:**

ORGAN
Level

1	2

| 3 | 4 |
| | |

| 5 |
| |

ORGAN SYSTEM
Level

| 1 | 2 |
| | |

Physiology and *Biological Disorder* **Fact Grids for Chapter 9:**

TISSUE
Level

1

ORGAN
Level

1

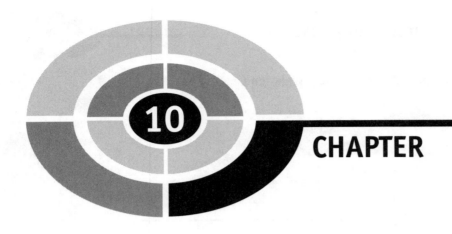

The Physiology of Blood: Helps It Walk with a "Lymph"

Chapter 9 discussed the physiology of the circulatory system in general. The blood was simply treated as the fluid "stuff" being pumped by the heart, throughout the body's vascular network.

But now we need to consider the *plasma* (**PLAZ**-muh) – the clear, watery, liquid "matter" (*plasm*) of the bloodstream! And the plasma has a close partner, the *lymph* (limf) or "clear spring water." The lymph is the clear, watery, intercellular material located between the cells of the *lymphatic* (lim-**FAT**-ik) *circulation*.

The blood (and its blood circulation) plus the lymph (and its lymphatic circulation) make a Dynamic Duo that function together to accomplish important things in the body.

The Blood–Lymph Physiological Relationship

Take a look at Figure 10.1. Here you see that the tiny *lymphatic capillaries* closely run by or shadow the blood capillaries, and that the lymph is, in fact, a *filtrate* (**FIL**-trait), or filtration product, of the blood. This filtration of blood occurs under the powerful pushing influence of the *capillary blood pressure*. This capillary BP pushes fluid and small particles out of the blood plasma, and across the selectively permeable membranes of the thin, flat endothelial cells lining both the blood and lymphatic capillaries. This pushed-out or filtered material from the plasma essentially becomes the lymph, which enters and flows through the lymphatic circulation.

To capsulize the blood versus lymph identities, we state two word equations:

Tissue 1

THE *BLOOD PLASMA* = THE CLEAR, FLUID, INTERCELLULAR "MATTER" CIRCULATING THROUGH THE *BLOOD* VESSELS

Tissue 1

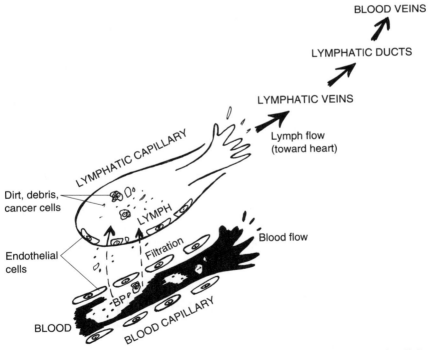

Fig. 10.1　The lymph: "clear water" (carrying dirt and contaminants) that "springs" from the bloodstream.

$$\begin{array}{ccc} & & \text{while} \\ \text{THE} & = & \text{THE CLEAR, FLUID, INTERCELLULAR } \textit{FILTRATE} \\ \textit{LYMPH} & & \text{(FILTRATION PRODUCT) OF THE BLOOD} \\ & & \text{PLASMA, CIRCULATING THROUGH THE} \\ & & \textit{LYMPHATIC } \text{VESSELS} \end{array}$$

Eventually, the tiny lymphatic capillaries merge to become larger *lymphatic veins*, and even huge *lymphatic ducts*. These lymphatic ducts eventually dump their load of lymph into several large blood veins. By this means, the lymph (which was originally formed by filtration out of the blood capillaries) is returned back to the bloodstream (by being dumped into several large blood veins) at the end of its circulation route.

Formed Elements in Blood and Lymph

Cell 1

An important idea in the anatomy of blood connective tissue is the concept of the *formed elements*. Now, let us not confuse these with the *chemical elements*! An element, in general, is (according to Early Latin) some *rudiment* (**ROO**-dih-ment) – a "beginning" or simple stage of something.

In anatomy, the formed elements are considered to be the whole cells, and the fragments of cells, found within the bloodstream (and to some degree, within the lymph as well). Thus, whole cells and fragments of cells are the *elements* or "rudiments" (the beginning parts) of blood connective tissue. And since each one is surrounded by a cell membrane, they also have a particular shape or *form*. Hence, the *formed elements* are cellular *elements* with a definite shape or *form*.

THE THREE GENERAL TYPES OF FORMED ELEMENTS

The three general types of formed elements created during hematopoiesis (Chapter 5) are the erythrocytes (red blood cells), leukocytes (white blood cells), and platelets or thrombocytes. (Go back and review Fig 5.8,B, if desired).

FORMED ELEMENTS IN THE LYMPH PLAY A ROLE IN IMMUNITY

Of the three major types of formed elements in the bloodstream, only one type – the leukocyte or white blood cell – is a regular member of the lymphatic circulation as well. The leukocytes are mainly known for the important functions they play in *immunity* (ih-**MYOO**-nih-tee). The word immunity translates to mean a "condition of" (*-ity*) "not serving" (*immun*) disease.

There are a number of different types of leukocytes. The specific type shown back in Figure 5.8 (B), you may recall, is called a *neutrophil* (**NEW**-troh-fil) or "lover" (*phil*) of "neutral" (*neutro*) chemical dyes. This means that the neutrophil is stained with neutral (neither acidic nor basic) dyes, in order to see its major structural characteristics through a light microscope. The nucleus of the neutrophil consists of two or three lobes connected by narrow segments. It also has tiny granules scattered throughout its cytoplasm. The neutrophil is the most common type of leukocyte. They usually make up more than 50% of all white blood cells found in the bloodstream.

Of course, neutrophils and the other types of leukocytes are too large to be simply filtered across the wall of the blood capillaries. But neutrophils and other WBCs can engage in *diapedesis* (**die**-uh-ped-**EE**-sis). This means that the white blood cells literally "leap through" the blood capillary by crawling in between the endothelial cells lining their walls. The crawling motion is technically called *ameboid* (ah-**ME**-boyd) *movements*, since this action resembles the movement of real amebas.

By diapedesis and ameboid movements, then, neutrophils that were once in the blood capillaries can cross over and enter the nearby lymphatic capillaries. Thus, we see them as formed elements in the lymph, as well as in the blood plasma.

Neutrophils engage in *phagocytosis* (**fag**-oh-sigh-**TOH**-sis), a "condition of" (*-osis*) "cell" (*cyt*) "eating" (*phag*). In particular, neutrophils are chemically attracted to sites of inflammation in the body. They leave the blood capillaries via ameboid movements and often enter the lymphatic circulation, which may contain invading bacteria. The neutrophils are the body's main bacteria-eaters, engulfing them by the thousands by the process of phagocytosis.

The lymphocytes

After the neutrophils, the second most common type of leukocyte are the *lymphocytes* (**LIMF**-oh-sights). The lymphocytes make up about 20–45% of the leukocytes circulating within the bloodstream. The lymphocytes have a distinct oval or lima bean-shaped nucleus that occupies most of the cell, with only a thin rim of cytoplasm. They lack visible granules in their cytoplasm.

The name lymphocyte means, of course, "lymph cell." This reflects the fact that most of the lymphocytes are not even found in the blood plasma!

Rather, most of them are found as formed elements within the lymph, either circulating within lymphatic vessels, or remaining within the *lymphatic organs*.

The two main categories of lymphocytes are the *B lymphocytes* and the *T lymphocytes*. The **B** or *bone marrow lymphocytes* are largely produced by the red bone marrow. The **T** or *thymic* (**THIGH**-mik) *lymphocytes*, in contrast, are named for the fact that they are often observed within the *thymus* (**THIGH**-mus) *gland*. The T lymphocytes and B lymphocytes often act together to fight foreign invaders by participating in the *antigen* (**AN**-tih-**jen**)–*antibody reaction* (which we will discuss shortly).

The monocytes

Finally, a third type of leukocyte (besides neutrophils and lymphocytes) is the *monocyte* (**MAHN**-oh-**sight**). The monocyte is a type of leukocyte named for its "single" (*mono-*), large, horseshoe-shaped nucleus. The monocytes frequently leave the bloodstream via diapedesis and ameboid movements, migrating to areas of infection or inflammation. Here they differentiate ("become different") by changing into *wandering macrophages* (**MAH**-kroh-**fah**-jes) or "big eaters."

Like the neutrophils, the monocytes (transformed into wandering macrophages) ingest large numbers of bacteria, cancer cells, and other foreign material by means of phagocytosis.

The Logical Lymph-Immune Connection

From our above discussion of the formed elements, we can conclude that most of the activity of the leukocytes in the protective *immune response* occurs within the *lymphatic* vessels or organs, *not* the *bloodstream*! Hence, whenever we talk about the *lymphatic system*, we are essentially talking about the *immune system* as well. It is only logical, therefore, for us to put these two organ systems together and describe them as a single *lymphatic–immune system*.

Capsulizing the preceding information, we have:

A

Organ System 1

LYMPHATIC SYSTEM	+	IMMUNE SYSTEM	=	LYMPHATIC–
(Lymphatic vessels and		(Antibodies and other		IMMUNE
organs containing lymph)		protectors from disease)		SYSTEM

"Dynamic Duo" Organs: Members of C-V and Lymphatic Systems

Now, the red bone marrow is considered part of the cardiovascular (blood circulatory) system, because it is one of the places where the formed elements of the blood are created. It is also the home of the B (bone marrow) lymphocytes.

But hematopoiesis occurs in other body organs as well. Consider, for example, the *spleen*. As shown in Figure 10.2, the spleen is one of the major lymphatic organs. And you also see the red bone marrow listed as a major lymphatic organ, too! Further, the spleen and red bone marrow help recycle old, beaten-up blood cells. Consequently, the spleen and red bone marrow can be considered "Dynamic Duo" (Batman and Robin) organs! They act as members of both the blood-producing cardiovascular system as well as the lymph-producing lymphatic system. And from our previous discussion, they are also part of the combined lymphatic–immune system.

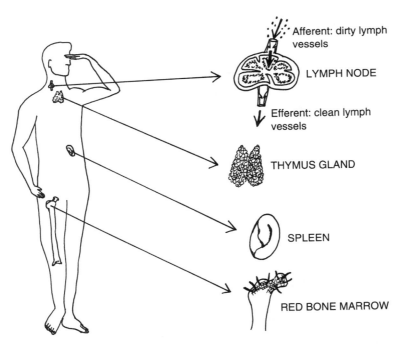

Afferent: dirty lymph vessels

LYMPH NODE

Efferent: clean lymph vessels

THYMUS GLAND

SPLEEN

RED BONE MARROW

Fig. 10.2 The major lymphatic organs.

Organ 1

We have:

"DYNAMIC DUO" ORGANS (MEMBERS OF BOTH THE C-V AND LYMPHATIC–IMMUNE SYSTEMS)	=	RED BONE MARROW + SPLEEN (C-V FUNCTIONS: Participate in hematopoiesis, destruction and recycling of old blood cells); LYMPHATIC–IMMUNE FUNCTIONS: Antigen–antibody reactions, phagocytosis of foreign invaders)

The Lymph Nodes and Thymus Gland

Two organs that are mainly lymphatic–immune in nature, without any direct connection to the cardiovascular system, are the *lymph nodes* and the *thymus* gland.

THE THYMUS GLAND

The thymus gland lies just deep to the sternum, in the middle of the chest. The thymus can be considered an endocrine gland of childhood and adolescence, because it reaches its maximum size at puberty, then progressively decreases in size with age. While it is still present, the thymus secretes the hormone *thymosin* (thigh-**MOH**-sin). This hormone stimulates the activity of the lymphocytes and other parts of the body's immune system. It also produces T (thymic) lymphocytes, which are sometimes just called *T cells*.

The thymus is thought to play an important role in the development of *immune competence* in youngsters – a growing ability to ward off various diseases. The disappearance of the thymus in adults may be related to the gradual decline of immune competence seen in older persons, thereby making them more susceptible to cancer and pneumonia.

THE LYMPH NODES

The most widespread lymphatic organs are the lymph nodes. The lymph nodes are a group of small, bean-like organs scattered in clusters in various parts of the body.

Afferent lymphatic vessels carry "dirty" lymph toward the lymph nodes. It is dirty in the sense that it often contains tiny particles of dirt, debris, or bacteria, which were filtered into the lymph from the blood capillaries. *Efferent lymphatic vessels* carry "clean" lymph away from the lymph nodes. And in "cleaning up" the dirty lymph that passes through them, the lymph nodes are essentially carrying out the basic processes of the *immune response*.

The Immune Response: Antibodies and Macrophages Attack the Antigens

Most cells carry chemical markers upon their surface membranes that uniquely identify them. Thus, cells in a particular human body have their own chemical surface markers that tag them as "self." Such cells are not attacked by the body's immune system. Foreign cells transplanted from some other body, or a bacterium or cancer cell, however, have different surface markers that tag them as "non-self." The general name for such surface markers is *antigens* (**AN**-tih-**jens**).

The word antigen means "produced" (*-gen*) "against" (*anti-*). This meaning reflects the fact that antigens are "non-self" marker proteins that label particular cells as foreign. Therefore, the antigens are foreign proteins that cause *antibodies* to be "produced against" them. Antibodies, in turn, are proteins produced by the body's immune system that attack and destroy foreign antigens. The overall process is called an *antigen–antibody reaction* (as mentioned earlier).

The antigen–antibody reaction comes about after a series of preceding steps (Figure 10.3). The first step is identification of a foreign cell and its surface antigen by a T (thymic) lymphocyte within the lymph nodes or some other lymphatic tissue. The T lymphocytes prowl around within the network of lymphatic vessels and act much like scouts. They send out a chemical signal whenever a foreign antigen is encountered.

The B (bone marrow) lymphocytes receive the chemical messages from the T lymphocyte scouts. The B lymphocytes then undergo a dramatic differentiation (process of becoming specialized or different). They transform into an entirely different cell type, called *plasma cells*. These new plasma cells have a prominent "clock face" nucleus when viewed through a compound light microscope. There is dark *chromatin* (kroh-**MAT**-in) visible – strands of DNA that have not yet coiled together to create chromosomes. These

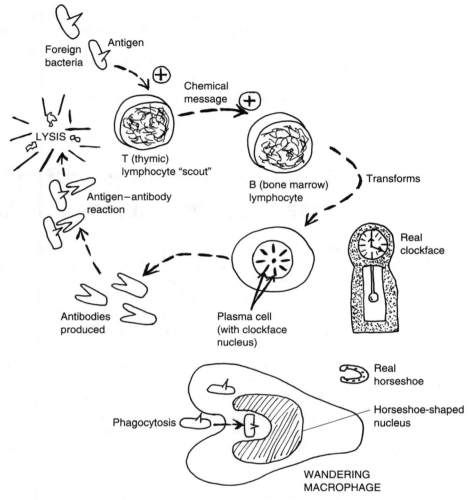

Fig. 10.3 The immune response.

chromatin fragments are arranged in a circular fashion around the edges of the nucleus, giving it a distinct "clock face" appearance.

It is the plasma cells that actually produce the antibodies. Once produced, the individual antibody molecules attach to the foreign antigens, like two pieces of a jigsaw puzzle fitted together. The result we have called an antigen–antibody reaction. When the antibody combines, it causes a lysis (breakdown) of the invading cell carrying the foreign antigen. Thus, millions of invading or abnormal cells (and their antigens) are efficiently ruptured and scattered into tiny pieces.

Moving nearby is a defensive army of neutrophils and larger wandering macrophages (representing transformed monocytes). The macrophages (and their smaller neutrophil allies) readily surround and engulf entire invading cells and their foreign antigens by means of phagocytosis (cell eating).

SUMMARY

Hence, by two major processes – antigen-antibody reactions and phagocytosis – an immune response has occurred. As its name indicates, the immune response produces a state of immunity from disease.

To verbally encapsule the immune response, we have:

| THE IMMUNE RESPONSE (STATE OF IMMUNITY FROM DISEASE) | = | ANTIGEN– ANTIBODY REACTIONS (Cause lysis of invading cells) | + | PHAGOCYTOSIS (Eating of invaders by macrophages) |

Cell 1

The ABO and Rh Antigen Systems

Our erythrocytes (like almost all of our other body cells) have chemical surface markers on their plasma membrane. But instead of being called *antigens*, they are called *agglutinogens* (ah-gloo-**TIN**-oh-jens). [**Study suggestion:** Observe that both the word antigen and the word agglutinogen end with the same suffix, -*gen*. Use this fact to help you remember that an agglutino*gen* is a type of anti*gen* found on the plasma membranes of red blood cells.]

The word agglutinogen literally means "a producer" (-*gen*) of "glueing or clumping" (*agglutin*). What is meant by this is made clear by a look at Figure 10.4. When two blood types are *incompatible*, the RBCs of the *blood donor* cannot be safely given to the *blood recipient*. The reason is that the blood recipient has *agglutinins* (a-**GLOO**-tin-ins) or chemical antibodies in its serum that attack the agglutinogens on the incompatible RBCs and chemically lyse (split) them. The exploded RBCs lose their hemoglobin, and their shriveled remnants – called *RBC "ghosts"* – then "clump" or *agglutinate* (ah-**GLOO**-tih-nayt) together into big, sticky balls. These big balls of RBC ghosts can block or *occlude* (ah-**KLOOD**) important blood vessels, thereby starving the tissue cells downstream.

Tissue 1

In short, we have a special case of the antigen–antibody reaction. Here, it is called a *transfusion reaction* between incompatible blood types. This is the

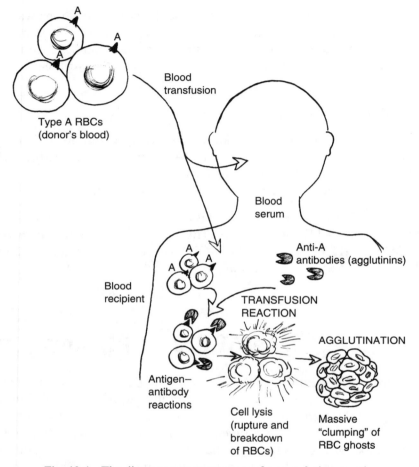

Fig. 10.4 The disastrous consequences of a transfusion reaction.

reason that a blood bank will carefully *type and cross-match* the blood between the recipient and a potential donor, to see if their blood types are *compatible* (capable of being transfused without causing RBC splitting and clumping).

THE ABO ANTIGEN SYSTEM

The most familiar group of agglutinogens on RBC membranes is the *ABO antigen system*. In this system, each RBC is assigned an ABO type based upon the chemical markers (agglutinogens or antigens) present upon its cell membrane. (Study Figure 10.5.) *Type A* red blood cells, for instance, have the

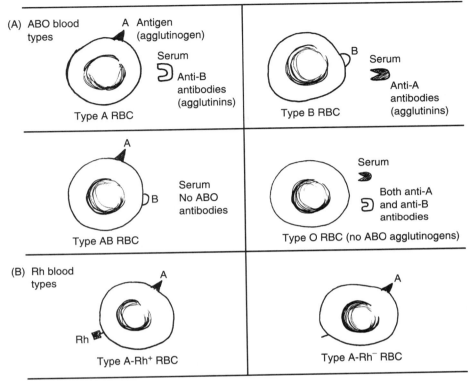

Fig. 10.5   The ABO and Rh antigen systems on red blood cells. (A) The ABO blood types. (B) An example of Rh-positive versus Rh-negative blood types.

A antigen present on their cell membrane. *Type B* has just the *B antigen, type AB* has both the A and B antigens on its membrane, but *Type O* is essentially a zero, having no AB antigens on its RBC surfaces.

The most important thing to consider in making a transfusion is the chemical marker (agglutinogen) present on the donor's RBCs, and the agglutinins (antibodies) in the recipient's blood serum. (This is because there are millions of agglutinins within the recipient's serum, which can easily cause agglutination or clumping together of donated RBCs.)

Consider, once again, Type A blood. In the serum are millions of *anti-B agglutinins or antibodies.* Therefore, if Type B blood is donated to a Type A person, the swarms of anti-B agglutinins in the Type A person's serum will massively destroy and agglutinate the donated Type B RBCs, causing a dangerous transfusion reaction. Therefore, Type A and Type B are considered incompatible blood types.

[**Study suggestion:** From a careful examination of Figure 10.5, why do you think that Type O is considered the "universal ABO donor"? Why is Type AB called the "universal ABO recipient"?]

THE RH FACTOR

In addition to the ABO antigen system, there are many others present on the RBC membrane. Most important among these are the *Rh factor or antigen*. The *Rh* name comes from the fact that this agglutinogen (antigen) was first observed on the RBCs of a *Rhesus* (**REE**-sus) monkey. Later, it was also found to occur on the RBCs of about 85% of all humans.

People who have the Rh factor on their RBCs are called *Rh⁺* or *Rh-positive*. The approximately 15% of people who lack the factor are called *Rh⁻* or *Rh-negative*. The Rh factor is usually reported in combination with a person's ABO antigen type. A person with both the *A* and *Rh antigens* present on each red blood cell, for example, is classified as *A positive*, or as *Type A, Rh positive*. If they have the *A* antigen but lack the *Rh* antigen then they are classified as *A negative*, or as *Type A, Rh negative*.

A person who is Type A, Rh negative, automatically contains anti-B agglutinins (antibodies) in the blood serum. But no anti-Rh antibodies or agglutinins are naturally present in the blood serum. Anti-Rh antibodies are produced by the immune system only some time *after* an Rh-negative person has somehow received a quantity of Rh⁺ RBCs. This can happen in the blood serum of an Rh-negative woman after she has given birth to an Rh-positive child. The reason is that some of the child's blood may have mixed with the mother's during the birth process, say from small scrapes or wounds on the child's head and the women's *genital* (**JEN**-ih-tal) area. If the mother's immune system has been "Rh-sensitized" in this way, then the RBCs of any subsequent Rh⁺ *fetus* (**FEE**-tus) in her womb may be attacked and destroyed by her anti-Rh antibodies.

The Process of Blood Clotting

The only type of formed element in the blood that we have not yet discussed in this chapter are the platelets or thrombocytes. The platelet name indicates their anatomy – shaped like "little plates." The thrombocyte name indicates their function – participation in *thrombus* (**THRAM**-bus) or "clot" (*thromb*) formation.

Tissue 2

HEMOSTASIS: A POSITIVE FEEDBACK PROCESS

The process of blood clotting results in *hemostasis* (hee-**MAHS**-tah-sis) – "a control of" (*-stasis*) "bleeding" (*hem*). [**Study suggestion:** Note how similar the spelling of *hemostasis* is compared to *homeostasis*. Do you still remember the literal translation of the word homeostasis? How does it differ from that of hemostasis?]

Blood clotting and hemostasis is a good example of a positive feedback system that is beneficial (and often life-saving) rather than harmful to the body. Figure 10.6 provides an illustration of the essential steps in this process.

First, the blood plasma in an undamaged vessel contains two important types of *clotting proteins*. They are named *fibrinogen* (feye-**BRIN**-oh-jen) and *prothrombin* (proh-**THRAM**-bin). Fibrinogen is very soluble (dissolvable) in blood plasma, so it will not settle out to produce a clot. And prothrombin is

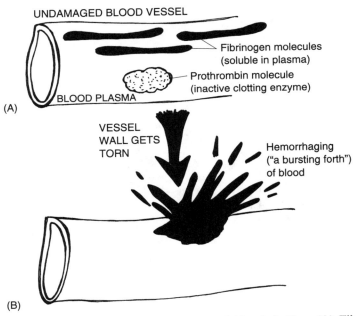

Fig. 10.6   Some of the major steps in hemostasis and blood clotting. (A) Fibrinogen and prothrombin within an undamaged blood vessel. (B) Hemorrhaging occurs after vessel injury. (C) Vascular spasm (powerful vasoconstriction) narrows injured vessel. (D) Platelet plug forms over torn collagen fibers in vessel wall. (E) Laying down of fibrin meshwork starts a positive feedback process between bigger and bigger thrombus (blood clot) size, and slower rate of blood loss.

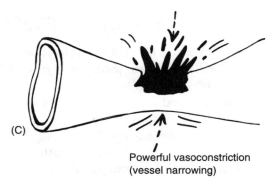

(C)

Powerful vasoconstriction
(vessel narrowing)

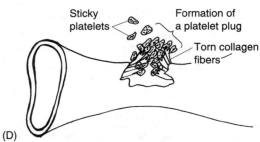

Sticky
platelets

Formation of
a platelet plug

Torn collagen
fibers

(D)

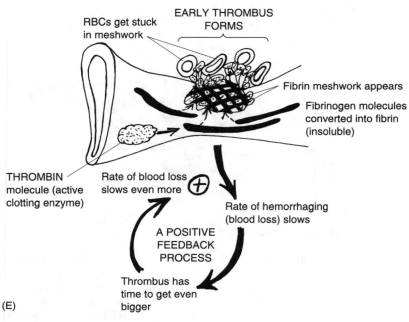

RBCs get stuck
in meshwork

EARLY THROMBUS
FORMS

Fibrin meshwork appears

Fibrinogen molecules
converted into fibrin
(insoluble)

THROMBIN
molecule (active
clotting enzyme)

Rate of blood loss
slows even more ⊕

Rate of hemorrhaging
(blood loss) slows

A POSITIVE
FEEDBACK
PROCESS

Thrombus has
time to get even
bigger

(E)

Fig. 10.6 (continued)

an inactive clotting enzyme which does not promote any clotting process. But these clotting proteins are always there, ready to be transformed into other chemicals that *do* participate in the clotting process!

Things change rapidly, then, whenever the wall of a blood vessel is torn and *hemorrhaging* (**HEM**-eh-rij-ing), or "a bursting forth" of bleeding, occurs. Here are some of the major steps leading to hemostasis (and complete stopping of hemorrhaging), after a vessel wall is injured.

Hemostasis step 1: Occurrence of powerful vasoconstriction (vascular spasms)

Right after a blood vessel is injured, it undergoes a powerful vasoconstriction (vessel narrowing). Because the vasoconstriction is so strong and sudden, it is often called a *vascular spasm*. This spasm and narrowing of the vessel lumen immediately slows the rate of blood loss, because less blood is now flowing through the constricted, injured vessel.

Hemostasis step 2: Formation of a platelet plug

When a vessel is cut open, the collagen fibers in its wall are torn. Platelets in the surrounding blood become sticky and attach to the shredded ends of the collagen fibers. As they pile up, a *platelet plug* is created that partially closes off the hole in the vessel wall.

Hemostasis step 3: Coagulation or blood clotting

After the platelet plug is formed, *coagulation* (koh-**ag**-you-**LAY**-shun) – the "process of" (*-tion*) "clotting" (*coagul*) – swings into full action.

Injury to the vessel wall creates a chemical called *prothrombin activator*. As its name indicates, prothrombin activator converts prothrombin (an *inactive* clotting enzyme) into an *active* clotting enzyme. Now the very word *prothrombin* literally means "before" (*pro-*) "thrombin." Therefore, prothrombin activator converts prothrombin into what comes after it – *thrombin*. Thrombin is an active clotting enzyme that really gets things started!

Thrombin acts upon fibrinogen, which we said was soluble or dissolvable within the bloodstream. Again translating a term, note that fibrinogen means "producer" (*-gen*) of "fibrin." Consequently, a new protein, *fibrin* (**FYE**-brin), appears.

Fibrin means "a fiber" (*fibr*) "substance" (*-in*). The reason for this name is that fibrin is a thin, whitish-colored, fiber-like protein that exists as slender filaments. Fibrin filaments are insoluble (not dissolvable) within the blood plasma. Hence, the fibrin filaments settle out of the blood and are deposited as a *fibrin meshwork* over the platelet plug.

More circulating platelets, as well as a few RBCs, get stuck in this fibrin meshwork. A sticky, jelly-like thrombus results. A positive feedback process now begins.

The bigger the clot (thrombus) gets, the slower that blood is lost. And the slower that blood is lost (hence allowing more time for clotting to occur), the bigger the clot or thrombus gets. This vicious cycle or vicious circle between ever-increasing size of the thrombus, and ever-slowing rate of hemorrhaging from the site of the wound, keeps continuing until a dramatic climax is finally reached – the complete sealing off of the hole in the vessel wall by a big, gooey thrombus! The rate of blood loss falls to zero, and a process of healing of the damaged vessel wall takes place, while the clot is holding back the blood like an effective beaver dam! (But this dam consists of a meshwork of fibrin mixed with platelets and RBCs, rather than a dam of interlacing logs, twigs, mud, and leaves.)

Quiz

Refer to the text in this chapter if necessary. A good score is at least 8 correct answers out of these 10 questions. The answers are listed in the back of the book.

1. The major force behind the filtration of lymph:
   - (a) Osmosis of water down its concentration gradient
   - (b) Simple diffusion of ions through pores
   - (c) Capillary blood pressure
   - (d) Local circuit currents

2. Formed elements represent:
   - (a) Cells and cell fragments
   - (b) The atoms of particular chemical elements with distinct shapes or forms
   - (c) Dissolved solutes of various sizes
   - (d) Cell organelles having distinct structures and functions

3. Play an important role in immunity:
    (a) Platelets
    (b) Leukocytes
    (c) Erythrocytes
    (d) Thrombocytes

4. The process whereby WBCs can "leap" or crawl "through" capillary walls:
    (a) Diapedesis
    (b) Phagocytosis
    (c) Filtration
    (d) Saltatory conduction

5. Often change into wandering macrophages:
    (a) Lymphocytes
    (b) Plasma cells
    (c) Monocytes
    (d) T cells

6. The spleen and red bone marrow are both distinct in that they:
    (a) Contain oxygen-rich blood
    (b) Belong to both the cardiovascular and lymphatic–immune systems
    (c) Are exactly the same structures as lymph nodes
    (d) Filter lymph, rather than consume it

7. Chemical surface markers on cell membranes that tag them as "non-self":
    (a) Hemoglobin molecules
    (b) Hemocytoblast fragments
    (c) Agglutinins
    (d) Antigens

8. Cells that actually produce antibodies:
    (a) Stem cells
    (b) Plasma cells
    (c) B lymphocytes
    (d) Neutrophils

9. The occurrence of a transfusion reaction indicates that:
    (a) An incompatible blood type has been donated to a recipient
    (b) Two ABO blood types have been properly cross-matched
    (c) No Rh factors are present
    (d) The immune response has become non-functional

10. Is laid down as a skeletal meshwork during clot formation:
    (a) Prothrombin
    (b) Myosin
    (c) Fibrin
    (d) Coenzyme Q

Body-Level Grids for Chapter 10

Several key body facts were tagged with numbered icons in the page margins of this chapter. Write a short summary of each of these key facts into a numbered cell or box within the appropriate *Body-Level Grid* that appears below.

Anatomy and *Biological Order* **Fact Grids for Chapter 10:**

A

CELL
Level

1

TISSUE
Level

1

ORGAN
Level

1

ORGAN SYSTEM
Level

1

Physiology and *Biological Order* Fact Grids for Chapter 10:

CELL
Level

1

TISSUE
Level

1	2

Physiology and *Biological Disorder* Fact Grids for Chapter 10:

TISSUE
Level

1

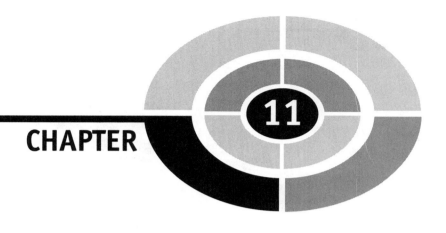

The Respiratory System: Makes You "Breathe Again"

Chapter 10 in *PHYSIOLOGY DEMYSTIFIED* outlined some pressures, volumes, and flows of two liquids. These were the blood and the lymph. Now Chapter 11 will address still another fluid substance – the air. And, as before, we will see how its physiology behaves under pressure, at certain volumes, and with certain rates of flow.

The basic structures of the respiratory system are thoroughly covered in *ANATOMY DEMYSTIFIED*. In this book, we will be focusing upon the lowermost tips of the so-called respiratory tree. These tips largely consist of the *bronchioles* (BRAHNG-kee-ohls) or "little bronchi". The bronchioles have a very high proportion of *circular smooth muscle* in their walls. (Consult Figure 11.1.)

A

Organ 1

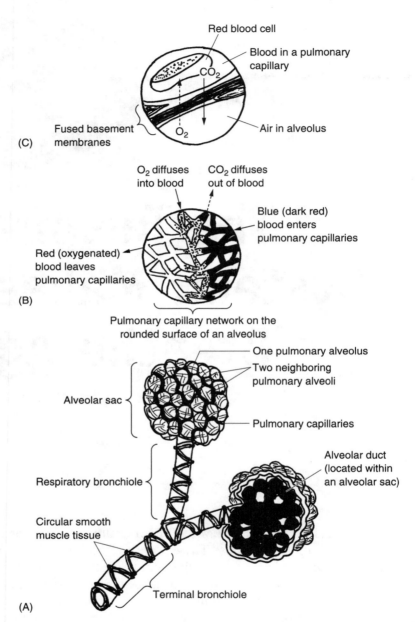

Fig. 11.1   A close-up view of the bronchioles and alveoli. (A) The final airways leading into the alveolar sacs. (B) A pulmonary capillary network changes its color from blue to red blood (due to addition of O_2 from the alveolus, and exit of CO_2 from the bloodstream. (C) Most highly magnified view: the actual respiratory membrane (air–blood barrier between an alveolus and pulmonary capillary).

BRONCHODILATION VERSUS BRONCHOCONSTRICTION

This high amount of circular smooth muscle (plus the lack of any cartilage horseshoes) allows the bronchioles to greatly widen (dilate), or greatly narrow (constrict), the diameter of their lumens. During exercise or severe stress, the *bronchiolar* (**brahng**-kee-**OH**-lar) smooth muscle relaxes, loosening the wall of the bronchioles. This results in *bronchodilation* (**brahng**-koh-dih-**LAY**-shun) – the "process of bronchial tube widening". The amount of frictional resistance or "rubbing" of the air molecules against the greatly enlarged tube walls is therefore reduced. Hence, air flows through the dilated bronchioles in greater volumes.

This greater volume of air going into and out of the lungs, of course, is of considerable help in supplying the tissue cells with greater amounts of O_2 to support their aerobic metabolism. It also allows greater quantities of CO_2 to flow away from the tissues as metabolic waste products. These greater amounts of air flowing through dilated bronchioles are largely the result of the increased activity of the sympathetic nerves, and of the adrenal medulla and its secretion of epinephrine (Chapter 8, Figure 8.3).

Now, let us consider rest, relaxation, and digestion. These are the body conditions when the *parasympathetic* (**PAIR**-uh-**SIM**-pah-**thet**-ik) *nerves* are dominant. (The sympathetic and parasympathetic nerves are discussed in more detail within the pages of *ANATOMY DEMYSTIFIED*.) In this situation, most of the tissue cells (except those lining the digestive tract) are relatively inactive. Hence, the circular smooth muscle in the walls of the bronchioles tends to contract with moderate force. The reason is that the parasympathetic nerves to the bronchiole wall stimulate the contraction of the circular smooth muscle. Like a noose lightly tightening around a cowboy's neck, the result is *bronchoconstriction* (**brahng**-koh-kahn-**STRIK**-shun) – "the process of narrowing of the bronchial tubes."

As a result of this bronchoconstriction, there is an increased frictional resistance to air flowing through the bronchioles, and therefore the volume of air moving through them goes down.

Remember:

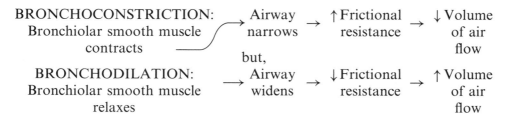

BRONCHOCONSTRICTION: → Airway → ↑Frictional → ↓Volume
Bronchiolar smooth muscle   narrows   resistance   of air
    contracts   flow
                             but,
BRONCHODILATION: → Airway → ↓Frictional → ↑Volume
Bronchiolar smooth muscle   widens   resistance   of air
    relaxes   flow

Organ 1

GETTING DOWN TO THE BUSINESS OF "REAL" RESPIRATION

There is a series of smaller and smaller bronchioles. These, of course, may either bronchoconstrict or bronchodilate. The smallest (and last) of them all is called the *respiratory bronchiole*. "Why is it called a *respiratory* bronchiole, Professor? None of the other bronchioles have this name."

That is correct, Baby Heinie. The reason for this name is that the respiratory bronchiole is the bronchiole leading into the *respiratory membrane*, where real *respiration* occurs! You will note from the extreme close-up (back in Figure 11.1) that the respiratory bronchiole is followed by only two more structures. These are the *alveolar* (al-**VEE**-oh-lar) *duct*, which leads into an *alveolar sac*.

"What's an alveolar sac, Professor Joe?" Well, it looks a lot like a bunch of grapes, doesn't it? In reality, however, each individual grape-like structure is really a *pulmonary alveolus* (al-**VEE**-oh-lus). Like a bunch of grapes hanging on a hollow stem, an alveolar sac, consisting of a bunch of pulmonary *alveoli* (al-**VEE**-oh-lie), hangs from a hollow alveolar duct.

"What's an alveolus, Professor?" The word alveolus means "little cavity." So, a *pulmonary* alveolus is a "little cavity" within the "lungs" (*pulmon*). The two lungs together, in fact, contain about 300 million pulmonary alveoli!

"Wow! They must really be *important*!" Yes, each alveolus is important. There are two main reasons:

1. The alveolus is the very end of the respiratory "tree," that is, the extensively branching series of tree-like bronchi and bronchioles.
2. The alveolus is the only structure in the entire so-called "respiratory" system where real respiration actually occurs!

Pulmonary ventilation not the same thing as pulmonary respiration

"But I thought this whole branching deal in the lungs was the respiratory system, which you said involved the process of breathing again and again!" I can see why you are somewhat confused by this unfortunate terminology, Baby Heinie. We badly need to distinguish pulmonary *respiration* from *pulmonary ventilation* (**ven**-tih-**LAY**-shun).

Pulmonary ventilation is the process of sucking air into the lungs, and blowing air out of the lungs. (This is a lot like a ventilation system for a large factory building, which both sucks clean air in, from the outdoors, and then blows dirty air out.) Therefore, when we are actually *breathing*, we are carrying out pulmonary *ventilation*.

"Okay, then what is pulmonary *respiration*?" What physiologists really mean by respiration is gas exchange between two or more body compartments. "*Which* compartments?" Well, take another look back at the real close-up in Figure 11.1. The two compartments involved in pulmonary respiration are the air within each alveolus, and the blood within each pulmonary capillary, lying right next to it.

"Is this area what you meant when you were talking about a respiratory membrane?" Yes, it is! Another name for the respiratory membrane is the *alveolar–capillary membrane*. Now, do you see in Figure 11.1 how the pulmonary capillaries form a webbing over the alveoli? And do you also observe that the first half of the pulmonary capillaries contains blue (dark) blood, while the second half undergoes a color change to red (lighter-colored) blood? This color change in the blood is due to pulmonary respiration.

"So, you mean that it is at the respiratory membrane, actually the alveolar–capillary membrane, where respiration really occurs?" Yes, this is the place where the wall of each alveolus is tightly fused against the wall of its pulmonary capillary. Each type of chamber is lined by only a single layer of endothelial cells. Thus, it is an excellent membrane for the diffusion of O_2 molecules out of the alveolus, into the pulmonary capillary, and for CO_2 molecules to diffuse in the opposite direction.

Organ system 1

"What is going on in the rest of the so-called respiratory system, Professor?" Well, the rest of this organ system should really be called the *ventilatory* (**VEN**-tih-lah-**tor**-ee) *system* (rather than the *respiratory* system), because its chief function is *ventilation* (not respiration)!

"But don't many of the bronchi, bronchioles, and all that, run right along blood vessels, too, so that they could exchange O_2 and CO_2 with them?" In theory, yes. But in practical reality, no. [**Study suggestion:** See if you can follow through on Professor Joe's line of reasoning. Explain why pulmonary respiration cannot occur across the walls of the trachea, bronchi, and bronchioles. Check your reasoning with a knowledgeable friend.]

Remember:

PULMONARY RESPIRATION	=	Gas exchange across the respiratory membrane (alveolar–capillary wall)
		while
PULMONARY VENTILATION	=	The sucking of air into the lungs, and the blowing of air out of the lungs

Organ 2

Ventilation: A Bulk Flow of Air

If pulmonary ventilation is the process by which we breathe, then it would be good to understand more about it. In particular, it would be good to understand the steps of inspiration (*sucking* air *into* the lungs), as well as those of *expiration* (**eks**-pir-**AY**-shun) or *blowing* air *out* of the lungs.

PULMONARY   =   INSPIRATION   +   EXPIRATION
VENTILATION   (*Sucking* air into   (*Blowing* air out of
                              lungs)   lungs)

VENTILATION AS A BULK FLOW OF AIR

Ventilation and the movement of air is quite similar to blood pressure and the movement of blood. As we learned in Chapter 9, blood flows down a blood pressure (BP) gradient, from a place where the BP is higher, toward another place where the BP is lower. In general, we can call this a *bulk flow* process. Bulk flow is the pressure-driven movement of some fluid substance (such as blood or air) from an area of greater pressure toward an area of lower pressure. In other words, bulk flow occurs down a pressure gradient.

For ventilation to occur, then, the problem becomes one of creating an *air pressure gradient* – a difference in air pressure – between the air in the atmosphere and the air within the alveoli (tiny lung air sacs). With such a gradient, there will be a bulk flow of air from the atmosphere and into the lung alveoli.

EQUAL GAS PRESSURES

There are two total gas pressures to consider. The *atmospheric pressure* is the pressure created by all of the gases in the atmosphere. (The atmosphere is the approximately 1-mile-thick blanket of air covering the surface of the Earth.) At sea level, the atmospheric pressure pushes with a total force of about 760 mmHg. The atmospheric pressure also pushes with this force upon the lips and nostrils. So when a person opens his mouth, the atmospheric pressure tends to push air down into his lung alveoli. (Consult Figure 11.2, A.)

Conversely, the *intra-alveolar* (**in**-trah-al-**VEE**-oh-lar) *pressure* is the total pressure exerted by all of the gas molecules within the alveoli. When a person opens his mouth, the intra-alveolar pressure tends to push air out of the alveoli, and out of the nose and mouth.

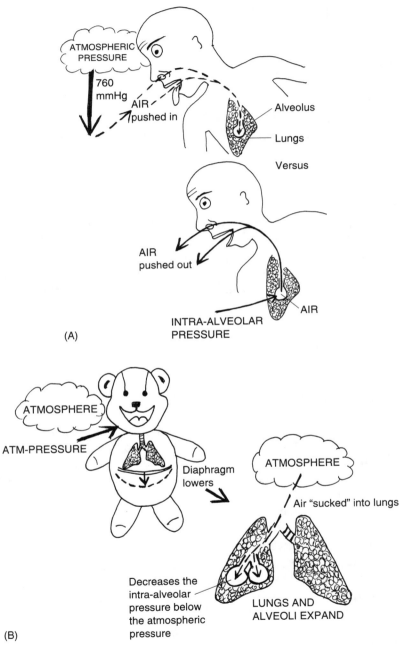

Fig. 11.2 Air pressures and inspiration in humans and in little bears. (A) Atmospheric pressure versus intra-alveolar pressure. (B) Negative pressure ("suction") breathing during inspiration.

The atmospheric pressure (tending to push air into the alveoli) and the intra-alveolar pressure (tending to push air out of the alveoli) are thus two opposing pressures. Between breaths, we have an equality of these two pressures. There is no air pressure gradient, and therefore, there is no bulk flow of air into or out of the lungs.

Summarizing:

Organ 3

| BETWEEN BREATHS: (No air pressure gradient exists: no bulk flow into or out of lungs) | ATMOSPHERIC PRESSURE (Tends to push air *into* the lungs) | = | INTRA-ALVEOLAR PRESSURE (Tends to push air *out* of the lungs) |

INSPIRATION AS NEGATIVE-PRESSURE BREATHING

We humans must create an air pressure gradient between the atmospheric pressure in the air surrounding us, and the intra-alveolar pressure within our lungs, if we are to achieve ventilation – a bulk flow of air either into, or out of, our lungs. We will trace the mechanism for creating inspiration, and then just reverse the process, for expiration.

Major steps in inspiration

(Study Figure 11.2, B.)

1. The *diaphragm* (**DIE**-uh-**fram**) or "barrier" muscle forming the floor of the chest cavity contracts. As the diaphragm contracts, it drops down in its position.
2. This makes the thoracic (chest) cavity larger at the bottom end, since its floor has been lowered.
3. Since the thoracic cavity becomes larger, the lungs follow suit and enlarge along with it. And because the lungs have enlarged, the millions of tiny alveoli become larger air sacs as well.
4. Because the alveoli are larger, their limited number of contained gas molecules are pushing against the walls of a much larger sac. Now, pressure is the amount of force exerted against a certain amount of area. As the area of the alveoli gets larger, then, the intra-alveolar pressure within them is distributed over a considerably greater area of their wall.
5. As a result, the intra-alveolar pressure does something "negative" – it falls below the atmospheric pressure. A pressure gradient is now

created between the atmospheric pressure and the intra-alveolar pressure. Like water being sucked down into a drain having lower pressure, the air is sucked down into the lung alveoli by inspiration.

After air has been *actively* sucked into the lung alveoli by the *active*, energy-requiring contraction of the diaphragm muscle, the diaphragm just quits contracting, and expiration occurs *passively* (without need of additional energy input). Like making a sigh of relief when you relax, this is how you basically exhale.

[**Study suggestion:** Examine Figure 11.2 B, and just reverse the direction of the events. You will then see for yourself the mechanism of passive expiration.]

Major Lung Volumes and Capacities

The previous section basically told us *how* we breathe (ventilate air). This section now reveals *how much* we breathe. Specifically, this section deals with the various *lung volumes* and *capacities*. We will now examine some of these major physiological parameters describing various aspects of pulmonary (lung) function.

BABY HEINIE THROWING SNOWBALLS

To be entertaining, let us view a picture that might be entitled *Baby Heinie Throwing Snowballs* (Figure 11.3). Since we can see his frosty breath, and we know he is getting rather excited and exercising, we can describe some of the lung volumes and capacities for an "average"-sized (but somewhat emotionally immature) male like Baby Heinie.

Tidal volume (TV)

Baby Heinie takes a single breath while he is resting. This breath is called the *tidal volume*, abbreviated as *TV*. The tidal volume (TV) is named for its resemblance to a real tide – the moving of waves of water back and forth upon the sand of a beach. (Or it can be visualized as the amount of frosty air going into, or out of, Baby Heinie's mouth with each of his resting breaths.) Technically speaking, the tidal volume is the amount of air exhaled after the person takes a normal resting inspiration. The TV amounts to about 500 ml (milliliters) in an average human adult.

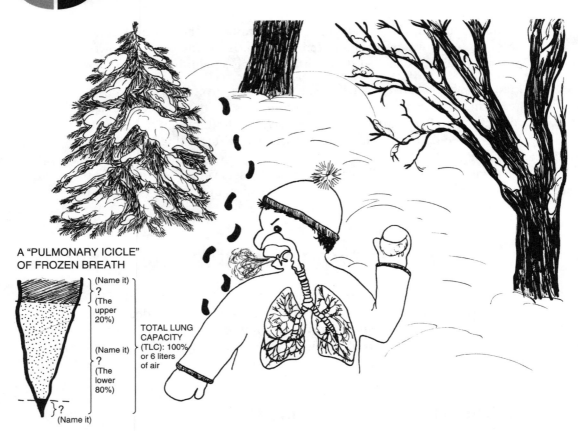

A "PULMONARY ICICLE"
OF FROZEN BREATH

(Name it)
? (The upper 20%)

(Name it)
? (The lower 80%)

}? (Name it)

TOTAL LUNG CAPACITY (TLC): 100% or 6 liters of air

Fig. 11.3 A kid throwing snowballs: help in visualizing some important lung volumes and capacities.

Vital capacity (VC)

Another important physiological parameter involving pulmonary function is the *vital capacity*, abbreviated as *VC*. *Vita* (**VEE**-tah) comes from the Latin and means "life." The vital capacity (VC) therefore represents a person's capacity for life. The VC is defined as the total amount of air that a person can inhale and exhale from normal, uncollapsed lungs. For a young male like Baby Heinie, the vital capacity is about 4,800 ml. (The vital capacity includes the 500 ml of air moved during the tidal volume.)

Residual volume (RV)

When Baby Heinie exhales, his alveoli normally do not totally collapse. If we picture each alveolus as a balloon, then the alveolus balloon only partially deflates during expiration. The balloon simply becomes smaller, rather than completely deflating. As a result, there is a *residual volume*, or *RV*. The residual volume (RV) is literally the residual or "left-over" volume that still remains within the alveoli, even after expiration has occurred. For BH (Baby Heinie), the residual volume is about 1,200 ml. This partial, residual inflation of the alveoli greatly reduces the amount of work required to completely re-inflate the lungs during the next cycle of inspiration.

Total lung capacity (TLC)

If we add both the vital capacity (VC) and the residual volume (RV) together, we obtain the *total lung capacity* (*TLC*):

TOTAL LUNG = VITAL CAPACITY + RESIDUAL VOLUME
CAPACITY (TLC)   (VC) (4,800 ml)   (RV) (1,200 ml)
6 liters (or 6,000 ml)

Organ 4

    The total lung capacity therefore represents the total amount of air that the lungs can possibly hold. For BH, the TLC is about 6 liters (or 6,000 ml). [**Study suggestion:** Figure 11.3 pictures the various lung volumes and capacities that we have been studying as the "frozen breath" of Baby Heinie, fused together to make a Giant Pulmonary Icicle, hanging down in space. The whole pulmonary icicle (100% of its length) represents the total lung capacity. The tiny pointed tip of 8.3% at the bottom of the icicle symbolizes what *specific* pulmonary parameter? The lower 80% of the icicle represents what *specific* pulmonary parameter? Finally, the upper 20% of the icicle denotes what *specific* pulmonary parameter? *Hint:* You just need to get out your calculator and do a little bit of simple math. When you are done, check your answers with that of a knowledgeable friend. Doesn't the bottom tip of any hanging icicle usually "melt" (get used up) first? How does this analogy apply to the stated lung volumes and capacities?]

Gas Transport: The Journey of O_2 and CO_2

Well, we now know *how* we breathe or ventilate, and we know *how much* we breathe or ventilate. The next question for the inquiring mind is, "Professor, how do we *carry* or *transport* what we breathe – the *respiratory gases*, O_2 and CO_2?" In other words, we need to address the topic that physiologists usually call *gas transport* through the bloodstream.

OXYHEMOGLOBIN AND THE RED BLOOD CELL

Certainly, the key issue for gas transport is the movement of oxygen (O_2) molecules through the bloodstream. When fresh O_2 diffuses into the blood across the respiratory membrane (alveolar–capillary wall), it does not mix or dissolve very well with the blood plasma itself. Instead, it diffuses into a nearby red blood cell (Figure 11.4, A).

Once inside the RBC, the O_2 combines with a *heme* (heem) *group* present within a *hemoglobin* (**HEE**-moh-**glohb**-in) molecule. Hemoglobin is a reddish-colored, "globe"-shaped (*glob*) "protein substance" (*-in*) found within the cytoplasm of the red "blood" (*hem*) cells. There are between 250 and 280 million *hemoglobin* (*Hb*) molecules present within a single RBC!

Each heme group is an iron-containing group that can combine with one oxygen molecule. Since there are four heme groups per hemoglobin molecule, obviously, one hemoglobin molecule can carry up to four O_2 molecules. [**Study suggestion:** From the previous information provided, estimate how many total O_2 molecules a *single* red blood cell can carry.]

When hemoglobin combines with O_2, the resulting combination is called *oxyhemoglobin* (**AHK**-see-**HEE**-moh-**glohb**-in). In this new combined molecule, the heme group turns a bright cherry red. This is the explanation, therefore, for the bright-red color of *oxygenated* (**AHK**-suh-jen-**AY**-ted) or oxygen-filled blood.

Using the chemical symbol of Hb for hemoglobin, we have

$$\text{Hb} + O_2 \quad \rightarrow \quad \text{HbO}_2$$
$$\text{(Hemoglobin)} \qquad \text{(Oxyhemoglobin)}$$

The O_2-loaded RBCs circulate to the tissue cells, which are actively consuming oxygen for their aerobic metabolism. The carried oxygen molecules then dissociate (split off) from the heme groups, and diffuse into the tissue cells.

BICARBONATE ION AND THE RED BLOOD CELL

The red blood cell performs a similarly important role in the transport of carbon dioxide (CO_2) molecules excreted during cellular respiration (the Krebs cycle and other aerobic processes, Chapter 4). CO_2, like O_2, is not very soluble (dissolvable) within the blood plasma. (Just about 7–10% of CO_2 is carried as gas bubbles dissolved in the blood plasma.) Hence, most of it ends up diffusing into the erythrocytes (Figure 11.4, B).

Here, the CO_2 is handled in two different ways:

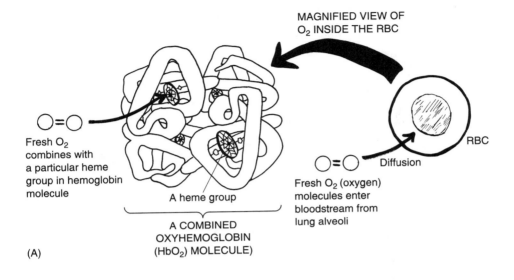

MAGNIFIED VIEW OF
O_2 INSIDE THE RBC

Fresh O_2
combines with
a particular heme
group in hemoglobin
molecule

A heme group

RBC

Diffusion

Fresh O_2 (oxygen)
molecules enter
bloodstream from
lung alveoli

A COMBINED
OXYHEMOGLOBIN
(HbO_2) MOLECULE)

(A)

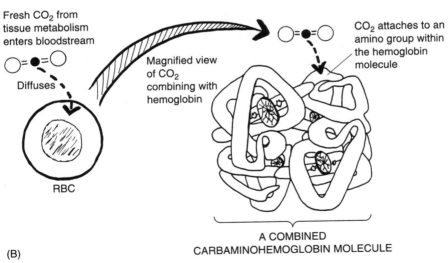

Fresh CO_2 from
tissue metabolism
enters bloodstream

Diffuses

Magnified view
of CO_2
combining with
hemoglobin

CO_2 attaches to an
amino group within
the hemoglobin
molecule

RBC

A COMBINED
CARBAMINOHEMOGLOBIN MOLECULE

(B)

Fig. 11.4   A brief look at gas transport. (A) Oxygen transport: Oxyhemoglobin and the red
blood cell. (B) Carbon dioxide transport on hemoglobin: creation of a carbaminohemoglobin molecule. (C) Carbon dioxide transported as bicarbonate ions in the
blood plasma.

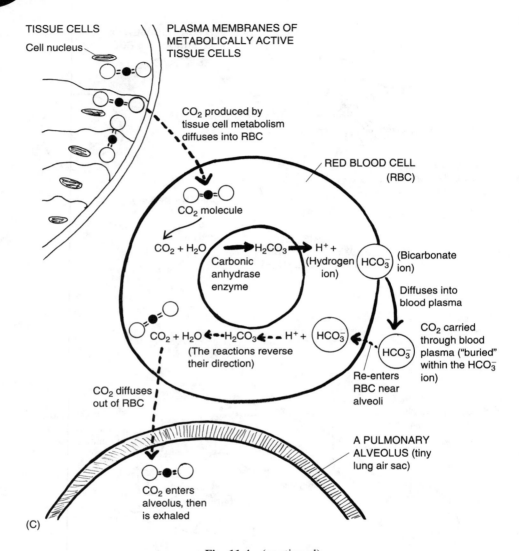

TISSUE CELLS
Cell nucleus

PLASMA MEMBRANES OF METABOLICALLY ACTIVE TISSUE CELLS

CO_2 produced by tissue cell metabolism diffuses into RBC

RED BLOOD CELL (RBC)

CO_2 molecule

$CO_2 + H_2O$ → H_2CO_3 → H^+ + (Hydrogen ion) HCO_3^- (Bicarbonate ion)

Carbonic anhydrase enzyme

Diffuses into blood plasma

$CO_2 + H_2O$ ← H_2CO_3 ← H^+ + HCO_3^- (The reactions reverse their direction)

HCO_3^-

Re-enters RBC near alveoli

CO_2 carried through blood plasma ("buried" within the HCO_3^- ion)

CO_2 diffuses out of RBC

A PULMONARY ALVEOLUS (tiny lung air sac)

CO_2 enters alveolus, then is exhaled

(C)

Fig. 11.4   (continued)

1. Combines with hemoglobin. The CO_2 molecules, once inside the RBC, combine with the *globin* (**GLOH**-bin), that is, the "globe"-shaped "protein" portion, of each hemoglobin molecule. (So, they do *not* compete with O_2 for binding with the heme groups.)

The combination of CO_2 with hemoglobin (Hb) therefore yields a new compound called *carbaminohemoglobin* (kar-**bam**-ih-noh-**hee**-moh-**GLOH**-bin). The prefix *carbamino-* (kar-**bam**-ih-noh) reflects the fact that "carbon dioxide" (*carb*) attaches to an *amino* (nitrogen-containing group) in the hemoglobin molecule. The chemical reaction is written as:

$$CO_2 + Hb \text{ (Hemoglobin)} \underset{\longrightarrow}{\longleftarrow} HbCO_2$$
(Carbon dioxide)   (Carbaminohemoglobin)

Approximately 20% of the blood CO_2 is carried within RBCs in the form of carbaminohemoglobin. As with O_2, when the red blood cells get near the pulmonary alveoli, the carried CO_2's dissociate (break off) from the Hb and diffuse into the alveoli (where most are soon exhaled). This backward reaction is symbolized by the arrows of the reaction in the equation between CO_2 and Hb, pointing in opposite directions.

2. Combines with water to create bicarbonate ion. By far the largest percentage (about 70%) of CO_2 diffuses into the RBC and simply combines with water. "How come the CO_2 doesn't combine with water *outside* of the red blood cell, Professor? Aren't there plenty of H_2O molecules available within the blood plasma?"

You are correct, Baby Heinie. There is a lot of water outside of the red blood cell, but only *inside* the RBC do you find the right *enzyme* that greatly *speeds up* the combining of CO_2 with H_2O, so that it occurs at a *significant* rate. This enzyme is called *carbonic* (kar-**BAHN**-ik) *anhydrase* (an-**HIGH**-drays). (Examine Figure 11.4, C.)

When CO_2 combines with H_2O (under the stimulating influence of carbonic anhydrase enzyme), the immediate product is *carbonic* (kar-**BAHN**-ik) *acid*, symbolized as H_2CO_3. The carbonic acid (H_2CO_3) molecule, like all acids (Chapter 3), is a hydrogen ion (H^+ ion) donor. Thus, the H_2CO_3 molecule quickly decomposes (breaks down) into both H^+ and HCO_3^- or *bicarbonate* (buy-**KAR**-boh-nayt) *ion*. We write:

F

Carbonic
anhydrase   So,

$$CO_2 + H_2O \underset{\longrightarrow}{\longleftarrow} H_2CO_3 \underset{\longleftarrow}{\longrightarrow} H^+ + HCO_3^-$$
(Carbon   (water)   (Carbonic   (Hydrogen   (Bicarbonate
dioxide)   acid)   ion)   ion)

Molecule 1

Now the bicarbonate (HCO_3^-) anion, unlike plain CO_2, *is* very soluble in the blood plasma! Therefore, the bicarbonate ions immediately diffuse out of the RBC, and mix with the blood plasma. (You can think of CO_2 as being "buried" or "housed" within each bicarbonate, HCO_3^- molecule.) The bicarbonate ions are quickly circulated through the blood plasma, and into the lungs. Once they reach the pulmonary capillaries, the HCO_3^- ions diffuse *back* into the RBC, and the reaction sequence that formed them in the first place is essentially *reversed*. This *re-creates* more CO_2, which then diffuses out of the

RBC, and finally into the pulmonary alveoli. Eventually, many of the CO_2 molecules diffusing into the alveoli are exhaled out of the body. (Hence, CO_2 is generally considered a body excretion – a metabolic waste product that is removed.)

SUMMARY OF GAS TRANSPORT

Let us now briefly recap the transport of oxygen and CO_2 within the bloodstream, as a couple of gas transport "rules."

Molecule 2

*The Blood **Oxygen** Rule:*
*Most blood oxygen is carried as **oxyhemoglobin** (HbO_2) within the red blood cells. Each O_2 molecule is attached to the **heme** (iron-containing group) within a hemoglobin (Hb) molecule.*

*The Blood **Carbon Dioxide** Rule:*
*Most blood carbon dioxide is "buried" or "housed" within a larger **bicarbonate** (HCO_3^-) ion, which circulates freely through the blood plasma and carries its CO_2 "cargo" long distances, into the lungs.*

Control of Respiration and Body Acid–Base Balance

The next logical question for us to ask is, "*Why* do we breathe? Without breathing, we quickly reach a state of severe *tissue hypoxia* (high-PAHKS-ee-ah). This is literally a 'condition of' (*-ia*) 'deficient or below normal' (*hyp*) blood 'oxygen' (*oxy*) levels. Doesn't it only make sense, therefore, that the major stimulus for inspiration would be a steep *drop* in the *blood O_2 concentration* toward its lower normal limit? After all, doesn't taking a nice deep breath replenish blood oxygen?"

Yes, that thinking makes good logical sense. But in reality it is not true. A steep drop in the blood O_2 concentration is too risky and dangerous to be the major stimulus for taking a normal, resting inspiration! The "wisdom of the body" does not allow us to wait to take our next inspiration until our oxygen-starved brains are practically on the verge of hypoxia! This approach is *foolish* (physiologically speaking)!

THE BODY ACID–RESPIRATION CONNECTION

"Gee, Professor Joe! If a drop in our blood oxygen level toward its lower normal limit *isn't* the major stimulus for our next resting, normal inspiration, then what is the stimulus?" The answer is direct. *The major stimulus for inspiration is a slight increase in the carbon dioxide (CO_2) concentration and acidity (H^+ ion) level within our bloodstream.*

Molecule 3

To see the mechanism, we need to go back to our old friend, the red blood cell, and peek within. Figure 11.5 (A) repeats most of the essential steps we talked about earlier regarding the blood transport of carbon dioxide. CO_2 produced from metabolism quickly reacts with H_2O within our erythrocytes. This reaction results in H_2CO_3 as its product. H_2CO_3, you will remember, is the chemical shorthand for carbonic acid. These carbonic acid molecules quickly break down into hydrogen (H^+) ions and bicarbonate (HCO_3^-) ions.

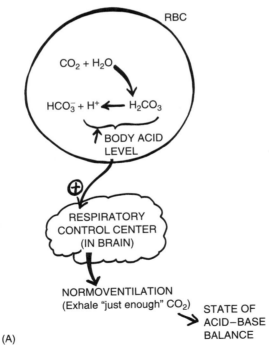

(A)

Fig. 11.5 Respiration and body acid–base balance. (A) Normoventilation achieves a healthy state of acid–base balance. (B) Hypoventilation and respiratory acidosis. (C) Hyperventilation and respiratory alkalosis.

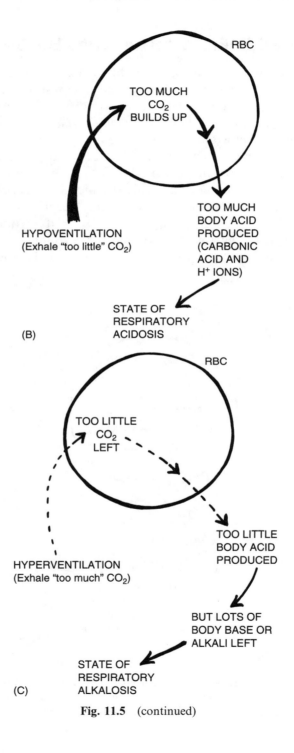

(B)

(C)

Fig. 11.5 (continued)

Hydrogen (H^+) ions and carbonic acid (H_2CO_3) are both classified as *body acids*. Recall that body acids are substances that either donate hydrogen ions (such as H_2CO_3 and HCl stomach acid) or that consist of hydrogen ions (such as a pool of H^+ ions, itself.) [**Study suggestion:** Chapter 3 discussed body acids, pH, and the concept of acid–base balance. It would be wise for you to review this information now, if you really don't remember much of it.]

Ventilation, of course, is the process of breathing (sucking air into the lungs, and blowing air out of the lungs). *Normoventilation* (**NOR**-moh-ven-tih-**LAY**-shun), therefore, involves breathing at a "normal" rate and depth for current metabolic conditions. In normoventilation, to be sure, the person inhales oxygen, but for the purpose of controlling breathing we are much more concerned with what happens during expiration. In this phase, the person exhales just enough CO_2 to prevent *respiratory acidosis*.

Respiratory acidosis is an "abnormal condition of" (-*osis*) too much body "acid," due to "respiratory" (breathing-related) causes. Because the person exhales just enough CO_2, there isn't time for too much carbon dioxide to accumulate within the RBCs and build up an excessive amount of either carbonic acid or hydrogen ions. Respiratory acidosis, therefore, is prevented by normoventilation, and a healthy state of *acid–base balance* is achieved.

HYPOVENTILATION AND RESPIRATORY ACIDOSIS

"Okay, Professor Joe," Baby Heinie scratches his chin while he thinks. "If normoventilation blows off just the right amount of CO_2 to prevent respiratory acidosis, then what if you exhale or blow off *too little* CO_2?"

Blowing off or exhaling too little CO_2 is what happens during *hypoventilation* (**HIGH**-poh-ven-tih-**LAY**-shun). By hypoventilation, we mean breathing at a "below-normal or deficient" (*hypo-*) rate and depth for current metabolic conditions. [**Study suggestion:** Try to name some specific situations where a person who was breathing at a *RR* (*respiratory rate*) suited for normal resting conditions, was no longer willing or able to continue ventilating sufficiently.]

A person hypoventilating long enough may fall into a state of respiratory acidosis (Figure 11.5, B). There is just too much CO_2 accumulating within the RBCs, due to a deficient rate of excreting (exhaling) them from the lungs. Not being blown off fast enough, too many CO_2 molecules react with H_2O, and too many carbonic acid molecules and hydrogen ions result. The state of acid–base balance is broken, and a disordered condition of acidosis reigns supreme. [**Study suggestion:** Go back to Chapter 3, if necessary, and look up the set-point value for blood pH, as well as its normal range.

During respiratory acidosis, below what specific pH value will the blood acidity likely fall?]

The brain can only function normally when the blood and CSF (cerebrospinal fluid) are in a state of acid–base balance. Hence, during respiratory acidosis the person may pass out, or even collapse into a coma!

HYPERVENTILATION AND RESPIRATORY ALKALOSIS

"Well, I don't want to pass out and die, or anything, Professor! So I promise that I won't act like a spoiled brat and hold my breath or *hypoventilate* (breathe deficiently), just to scare my Mom into buying me more expensive toys!" Good for you, Baby Heinie! "But what if I throw a tantrum and suddenly start kicking and screaming and bawling uncontrollably?"

In this case, Baby Heinie, you will be unwisely engaging in *hyperventilation* (**HIGH**-per-ven-tih-**LAY**-shun). Hyperventilation can often be seen in some emotionally overwrought and hysterically crying persons. In this situation, the RR and depth are at an "above-normal or excessive" (*hyper-*) level for current metabolic conditions (see Figure 11.6, C).

Therefore, the hyperventilating person blows off or exhales too much CO_2 from the body. Not enough CO_2 is left to react with water inside of the red blood cells. And there is not enough carbonic acid or hydrogen ion produced, because of this fact.

The resulting state is called *respiratory alkalosis* (**AL**-kah-**LOH**-sis). This is the technical term for an "abnormal condition of" (*-osis*) not enough body acid, or too much base or alkali, due to respiratory (breathing-related) causes. Here, of course, we have pointed out *emotional hyperventilation* as one potential cause of respiratory alkalosis.

If a person hyperventilates for too long, body acid levels fall way below their normal range, and a state of alkalosis follows. The person may well become dizzy and pass out. [**Study suggestion:** Above what specific pH value would the blood pH have to rise, for the person to be classified as suffering from respiratory alkalosis?]

Organ 5

RESPIRATION ACID–BASE SUMMARY

To capsulize, we can say that:

NORMOVENTILATION = Blow off = A state of acid–base (pH)
(Breathing at a normal   just enough   balance
rate and depth) CO_2

HYPOVENTILATION = Blowing off = A state of respiratory
(Breathing at a deficient   too little   acidosis (Too much body
rate and depth) CO_2 acid build-up)

HYPERVENTILATION = Blowing off = A state of respiratory
(Breathing at an excessive   too much   alkalosis (Too much body
rate and depth) CO_2 base or alkali, or not
                                         enough acid)

Organ 1

　　When a person is stimulated to inhale by a slight rise in blood acidity, then, he or she also exhales just enough CO_2 during expiration, such that a healthy state of acid–base balance is maintained. And, as a significant bonus, enough oxygen is delivered to the brain and other vital organs during inspiration to prevent tissue hypoxia. A normal pattern of aerobic functioning of the entire body is hopefully maintained.

Quiz

Refer to the text in this chapter if necessary. A good score is at least 8 correct answers out of these 10 questions. The answers are listed in the back of the book.

1. The first part of the respiratory system normally reached by air when it is inhaled:
   (a) Pulmonary alveoli
   (b) Smaller bronchi and bronchioles
   (c) Larynx
   (d) Nasal and oral cavities

2. The bronchioles are special in that they have the ability to:
   (a) Completely collapse under pressure
   (b) Transport blood as well as air
   (c) Rigidly support the chest wall
   (d) Dramatically dilate or constrict their lumens

3. The location where "real" respiration occurs:
   (a) Alveolar–capillary membrane
   (b) Plasma membrane of the RBC
   (c) Inner wall of the trachea
   (d) Lumen of the pharynx

4. The terminal grape-like clusters at the very distal end of the respiratory tree:
   (a) Respiratory bronchioles
   (b) Alveolar sacs
   (c) Horseshoes of cartilage
   (d) Dense fibrous connectors

5. Pulmonary ventilation differs from pulmonary respiration in that it:
   (a) Consists of gas exchange between air and the tracheal lumen
   (b) Involves movement through all of the airways
   (c) Is the blowing of air into the lungs, followed by the sucking of air out of the lungs
   (d) Only occurs in children

6. When the atmospheric pressure equals the intra-alveolar pressure, then:
   (a) No pressure gradient for bulk flow of air exists
   (b) Air will still move into the lungs, but slowly
   (c) Both CO_2 and O_2 will rush out of the lungs very forcefully
   (d) The diaphragm is probably contracting vigorously

7. Defined as the amount of air exhaled after a person takes a normal resting inspiration:
   (a) Residual volume
   (b) TLC
   (c) Vital capacity
   (d) Tidal volume

8. Most O_2 molecules are carried through the bloodstream by:
   (a) Combining with plasma proteins
   (b) Dissolving as gas bubbles
   (c) Forming oxyhemoglobin
   (d) Carbaminohemoglobin

9. Carbonic anhydrase is:
   (a) An enzyme that speeds up the catabolism of glucose
   (b) The major catalyst within RBCs for the reaction of CO_2 with water
   (c) A form of high-energy compound useful for kinetic force
   (d) Mainly responsible for carrying CO_2 through the bloodstream

10. The major stimulus for normal, resting inspiration:
    (a) A steep decline in the blood O_2 level
    (b) Paralysis of the diaphragm muscle

(c) An out-of-control rise in blood acidity toward acidosis

(d) A slight rise in the blood or CSF H^+ ion and CO_2 levels

Body-Level Grids for Chapter 11

Several key body facts were tagged with numbered icons in the page margins of this chapter. Write a short summary of each of these key facts into a numbered cell or box within the appropriate *Body-Level Grid* that appears below.

Anatomy and *Biological Order* **Fact Grids for Chapter 11:**

A

ORGAN
Level

1

Physiology and *Biological Order* **Fact Grids for Chapter 11:**

P

ORGAN
Level

1	2
3	4

5

ORGAN SYSTEM
Level

1

Physiology and *Biological Disorder* Fact Grids for Chapter 11:

ORGAN
Level

1

Function and *Biological Order* Fact Grids for Chapter 11:

MOLECULE
Level

1	2
3	

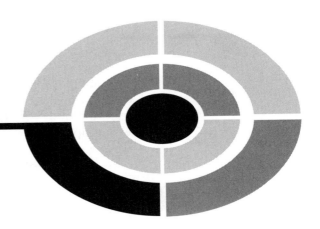

Test: Part 5

DO NOT REFER TO THE TEXT WHEN TAKING THIS TEST. A good score is at least 18 (out of 25 questions) correct. Answers are in the back of the book. It's best to have a friend check your score the first time, so you won't memorize the answers if you want to take the test again.

1. The circulatory system gets its name from the fact that:
   (a) Blood vessels circulate the blood back and forth, in a straight line
   (b) The spleen and liver are included in the system
   (c) The blood travels in a circular route, to and from the heart
   (d) Hematopoiesis occurs in some of its members
   (e) It follows a "circuitous" route that no has ever been able to trace!

2. The left heart is known as the pump for the:
   (a) Lungs and gall bladder
   (b) Pulmonary circulation
   (c) CPA
   (d) Systemic circulation
   (e) Pulmonary arterioles

3. The S-A node plays the principal role in:
   (a) Excitation of the cardiac muscle
   (b) Carrying blood from the atria
   (c) Messing up EKGs!
   (d) T-wave inversion
   (e) Prolapse of the mitral valve

4. The _____ "raises or lifts" blood "up" out of the left ventricle:
   (a) Aortic arch
   (b) Common pulmonary artery
   (c) Myocardium
   (d) Brachial artery
   (e) Superior vena cava

5. The contracting and emptying phase of both atria and both ventricles:
   (a) Diastole
   (b) Functional syncytium
   (c) Diastolic BP
   (d) SV
   (e) Systole

6. One complete sequence of excitation, contraction, and relaxation of all four chambers of the heart:
   (a) "Lubb"
   (b) Diastole
   (c) Closure of both semilunar valves
   (d) Snapping shut of both A-V valves
   (e) Cardiac Cycle

7. The "dup" sound generally indicates the event of:
   (a) Opening of both atrioventricular valves
   (b) Closure of both semilunar valves at the beginning of ventricular diastole
   (c) The middle of the resting period for all heart chambers
   (d) Closure of the right A-V valve, but opening of the left
   (e) Closure of the left S-L valve, but opening of the right

8. The P wave in the electrocardiogram reflects:
   (a) Depolarization of both atria
   (b) Repolarization of the ventricles
   (c) Hyperpolarization of the sino-atrial node
   (d) Loss of myocardial contractility
   (e) Ventricular fibrillation

9. Stroke volume × _____ = CO
   (a) TPR
   (b) HR
   (c) QRS
   (d) RR
   (e) TV

10. Portion of the BP that can be considered residual or "left-over" blood pressure:
    (a) Diastolic
    (b) Mean
    (c) Systolic
    (d) Acidolic
    (e) Proximal

11. Represents a clear, watery filtrate of the blood plasma:
    (a) Venous blood
    (b) Lymph
    (c) Kickapoo juice
    (d) Formed elements
    (e) Chyme

12. Neutrophils help immune activity mainly by carrying out:
    (a) Antibody synthesis
    (b) Blood clotting
    (c) Phagocytosis
    (d) Hematopoiesis
    (e) Hemostasis

13. Formed elements that appear only in blood plasma (not the lymph):
    (a) Plasma proteins
    (b) RBCs and WBCs
    (c) RBCs and platelets
    (d) Thrombocytes
    (e) T-cells

14. Antigens present on cell membranes:
    (a) Break off and soon circulate throughout the bloodstream
    (b) Often label particular cells as foreign or "non-self"
    (c) Discourage powerful responses from the immune system
    (d) Destroy all antibodies that attack them
    (e) Act as hosts for viruses

15. An agglutinogen is just a special type of:
    (a) Antigen present on the plasma membrane of some erythrocytes
    (b) Clotter of thin blood
    (c) Enzyme present within all leukocytes
    (d) Antibody moving through the blood plasma
    (e) Giant wandering macrophage

16. Severe tissue ischemia or a state of hypoperfusion suggests that:
    (a) The CO is very high and effective
    (b) The heart may have gone into cardiac arrest
    (c) The SV has increased considerably
    (d) Arterioles have vasodilated
    (e) Venous sinuses are overfilled

17. Cerebrovascular accident (stroke) represents a:
    (a) Negative feedback relationship between blood pressure and heart rate
    (b) Full-fledged "heart attack"
    (c) Ruptured aneurysm of the cerebral blood vessels
    (d) Homeostasis of blood pressure and blood flow
    (e) A random event having no cause or pattern

18. Type O blood:
    (a) The universal ABO donor
    (b) Contains only anti-A agglutinins (antibodies) in its serum
    (c) The universal ABO recipient
    (d) Has clotting factors for all formed elements
    (e) Contains only anti-B agglutinins (antibodies) in its serum

19. In a group of 100 people chosen at random, ___% are likely to have Rh-positive blood:
    (a) 10
    (b) 15
    (c) 50
    (d) 70
    (e) 85

20. Blood clotting is part of the whole process of:
    (a) Hemostasis
    (b) Leukocytosis
    (c) Homeostasis
    (d) Hematopoiesis
    (e) Vascularization

21. The protein laid down in a torn vessel wall as a meshwork of thin filaments:
    (a) Prothrombin
    (b) Fibrin
    (c) Thrombin
    (d) Fibrinogen
    (e) Hemoglobin

22. Flips up and down over the opening of the larynx:
    (a) Pharynx
    (b) Vocal cords
    (c) Endothelium
    (d) Epiglottis
    (e) Nasal cavity

23. Mainly responsible for bronchoconstriction and bronchodilation:
    (a) Alveoli
    (b) Bronchioles
    (c) Primary bronchi
    (d) Trachea
    (e) Alveolar sacs

24. Pulmonary ventilation = Inspiration + _____:
    (a) Perspiration
    (b) Inhibition
    (c) Respiration
    (d) Expiration
    (e) Titillation

25. The total amount of air a person can inhale and exhale from normal, uncollapsed lungs:
    (a) Tidal volume
    (b) Residual air
    (c) Diastolic BP
    (d) Total lung capacity
    (e) Vital capacity

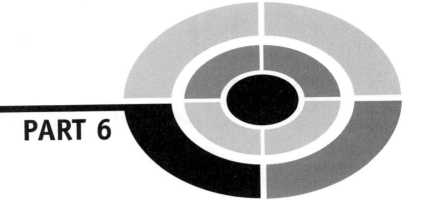

PART 6

Physiology from "The Land Down Under": The Digestive and Genitourinary Systems

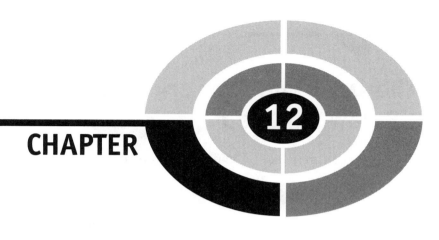

CHAPTER 12

The Digestive System: A Miraculous "Grinder"!

Well, the shining hero of Chapter 9 (Circulatory System), of course, was the heart. And the twin heroines of Chapter 11 (Respiratory System) were certainly the lungs. The "heart" (*cardi*) and the "lungs" (*pulmon*) have a very close *cardiopulmonary* (**kar**-dee-oh-**PULL**-moh-ner-ee) *relationship*. This is true anatomically, as well as physiologically, since both the heart and lungs are close neighbors within the thoracic cavity, which we can call "The Land Above the Belt."

The Dark and Mysterious Organ Systems "Below Our Belt"

In this, the last major section of *PHYSIOLOGY DEMYSTIFIED*, we must be brave. "Why is *that*, Professor?" We must be brave, Baby Heinie, because we are going to visit the Dark and Mysterious "Land Down Under"! By this I simply mean that we are going to visit two major organ systems that largely lie "Below the Belt." Specifically, these are the *digestive system*, plus the combined *genitourinary* (**JEN**-ih-toh-**ur**-ih-**nair**-ee) *system*. They both lie mostly below the diaphragm muscle, as close neighbors within the *abdominal* (ab-**DAHM**-ih-nal) or *abdominopelvic* (ab-**dahm**-ih-noh-**PEL**-vik) *cavity*.

Organ System 1

TWO ORGAN SYSTEMS   =   DIGESTIVE   +   GENITOURINARY
   "DOWN UNDER"   SYSTEM   SYSTEM

"How come they are so Dark and Mysterious, Professor? Our heart and lungs didn't seem to have any problems like that!" You just made a very interesting observation, young man! Isn't it strange that no one seems to get embarrassed about their heart pumping blood and their lungs pumping air, but many people seem really touchy and even ashamed about perfectly *normal* physiological processes going on in the "Land Down Under"!

Modern Human *Feces*, or Ancient Roman *Fasces*? Our "Dregs" Are All Packaged into "Bundles"

"Why is that, Professor?" Perhaps one reason is that certain "unmentionable" or "disturbing" things are coming *out* of the body, in the "Land Down Under." Specifically, for the digestive tube we have *feces* (**FEE**-seez) or "dregs" coming out of the *anus* (**AY**-nus). Medically speaking, the anus is the muscular "ring" (*an*) through which we *defecate* (**DEF**-eh-**kayt**) or "remove" (*de-*) our "dregs" (*fec*).

We can better understand this whole situation if we look at a simple summary diagram of the entire digestive tube (Figure 12.1).

The digestive system or tract is basically a long tube that begins with the "mouth" or *oral cavity*, and ends at the anus. (*NOTE*: The basic structure of

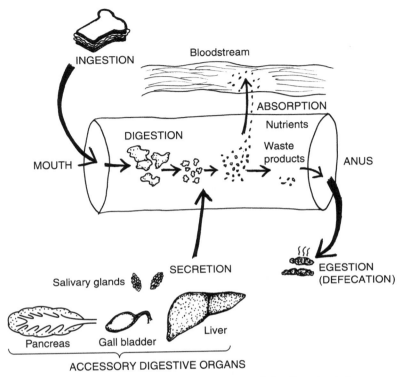

Fig. 12.1   A simple functional diagram of the digestive tube.

the digestive tube is clearly described within the pages of *ANATOMY DEMYSTIFIED.*) *Ingestion* (**in-JES**-shun) is the process of carrying food "into" (*in-*) the digestive tube through the oral cavity. *Egestion* is the process of carrying the feces "out" (*e-*) through the anus. And *digestion* is the process of "dividing or dissolving" the food into smaller pieces, after it has been *ingested* (in-**JES**-ted).

As a practical matter, then, the entire digestive tube is one long food "grinder," because it divides or dissolves eaten food into much smaller fragments.

THE LOWER GI TRACT FORMS THE FECES

"What do you mean by the lower GI tract?" Well, GI is an abbreviation for *gastrointestinal* (**gas**-troh-in-**TES**-tih-nal) – "pertaining to" (*-al*) the "stomach" (*gastr*) and the "intestines" (*intestin*). By the *lower* GI tract, therefore, we fundamentally mean the *inferior* portion of the digestive tube, consisting of everything from the *stomach* down through the *intestines* and into the *rectum* (**REK**-tum) – a "straight" (*rect*) tube leading into the anus. Summarizing, we have:

A

LOWER GI TRACT	=	INFERIOR PORTION OF DIGESTIVE TUBE	=	STOMACH + INTESTINES + RECTUM + ANUS

Organ Sysem 2

"But what exactly are the *intestines*, Professor?" The intestines are the *bowels* – two long, hollow, cylinder-shaped tubes, attached end to end, that extend below the stomach and look like a curled-up string of "little sausages."

The first 20 feet or so of the bowels are the *small intestine*, which resembles a tightly folded string of "small"-diameter sausages. This is followed by approximately 6 feet of the *large intestine* or *colon* (**KOH**-lun). The large intestine (colon), of course, resembles a kinked linkage of "large" diameter sausages.

A

INTESTINES (BOWELS OR "LITTLE SAUSAGES")	=	SMALL INTESTINE	+	LARGE INTESTINE (COLON)

Organ 1

"What does all this have to do with the feces or dregs?" Well, the intestines (bowels), especially the large intestine, absorbs lots of water and salt from the *chyme* (kighm). Chyme is a thick, soupy mass of partially digested material that is formed in the stomach, and then continues throughout most of the lower GI tract. The liquid, soupy chyme is progressively dried out as it slowly moves through the long, kinked sections of the large intestine. More and more water is absorbed into the bloodstream, until cylinder-shaped, semi-solid stools or feces are created.

"So the chyme is like a loose, watery mud in the stomach and small intestines? But then it gets dried out into caked mud or dregs, shaped like cylinders or sausages. And the reason is that the cylinder-shaped wall of the large intestine gradually presses or molds the chyme within its lumen to match the outline of its shape, right?" Correct!

ARE THEY FASCES, OR JUST "BUNDLES" OF FECES?

Even more fascinating is the related history behind this connection. Look carefully at the spelling and pronunciation of these two words: *fasces* (**FAS**-eez) and *feces* (**FEE**-seez). We have already noted that feces means "dregs" or "mud." We have added to this the fact that feces are generally shaped like sausages or cylinders. "Don't they also look like thick sticks, or maybe, bundles of thick brown sticks?"

This is a correct historical observation, Baby Heinie. During the Glory Days of Ancient Rome, high authorites would carry fasces (**FAS**-eez) – literally, "bundles" of cylinder-shaped, brown wooden sticks – over one shoulder. The bundles also contained a metal ax, with its sharp blade projecting out of the bundle (Figure 12.2).

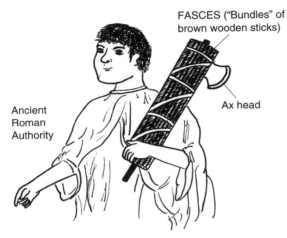

Fig. 12.2   The fascinating saga of the Ancient Roman fasces ("bundles").

These fasces were nothing to laugh about (if you valued your life!), because they were carried as symbols of high authority. How ironic it is, then, that the fasces or "bundles" of brown sticks carried by Roman officials to indicate high authority and imperial dignity, would eventually disappear and be replaced by another Latin-based word with just two different letters in it – the feces – which are bundles of dregs or human excrement!

Now, we also agreed that the feces were usually greenish-brown in color, like modern mud or the Ancient Roman fasces (bundles of brown sticks). But, surprisingly, if nothing is added to them from *outside* of the digestive tube, the feces will be defecated as a dull slate-gray in color. "Then, how do they go from slate-gray to *brown*?"

Enter the Accessory Digestive Organs

To answer this question, we will have to consider another major aspect of the physiology of the digestive tube that we have not yet discussed. This function is *secretion*. We have previously described secretion as being the principal action of the glands (Chapter 8). In the present case, secretion can be defined as the release of some useful substance into the lumen of the digestive tube.

"*Where* do these secretions come from, Professor?" There are some *goblet cells* lining the digestive tube wall, which secrete a sticky *mucus* into the lumen. Other gland cells also populate the tube wall, depending upon what area of the alimentary canal we are considering.

However, the major group of secreting structures are collectively called the *accessory digestive organs*. The accessory digestive organs are defined as organs that are attached to the side of the digestive tube as "accessories" (add-ons). They add useful secretions to the lumen, but they are not actually members of the digestive tube itself, because no food, chyme, or feces passes through them.

Way back in Figure 12.1, the accessory digestive organs were shown. These are the *salivary* (**SAH**-lih-**vair**-ee) *glands*, pancreas, liver, and *gall bladder* or *cholecyst* (**KOH**-luh-*sist*). All of these accessory digestive organs add useful secretions to the tube lumen.

Capsulizing all previous information on general digestive tube physiology, we state this word equation:

$$\begin{array}{ll} \text{GENERAL PHYSIOLOGY} & = & \text{INGESTION + SECRETION +} \\ \text{OF DIGESTIVE TUBE} & & \text{DIGESTION + ABSORPTION +} \\ & & \text{EGESTION (DEFECATION)} \end{array}$$

Organ System 1

Squirting Bile into the Digestive Tube: A Partnership Between the Liver and Gall Bladder

Let us now get back to the earlier question about what is added from the outside to color the slate-gray feces a greenish-brown color. Obviously, something is being secreted from one or more of the accessory digestive organs that colors the chyme. That something is *bile*. Bile is a brownish-green detergent substance. It is secreted by the liver, stored and released by the gall bladder, and then carried into the small intestine by the *common bile duct*. (Consult Figure 12.3.) The stools are brown, then, because brownish-green bile is added to the slate-gray chyme inside of the bowel.

The Ancient Greek word root for "bile" is *chole* (**KOH**-lee). [**Study suggestion:** Using the same "pertaining to" suffix as found in the word *physiologic*,

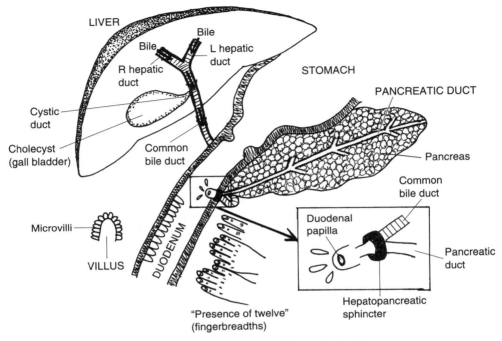

Fig. 12.3 A busy place for bile under the liver.

and the word root *chole*, see if you can write out a single word on a piece of paper, right now, that literally "pertains to bile." Check your answer with the actual word, which will now appear in the text.]

So, a word like *choleic* (koh-**LEE**-ik) therefore "pertains to bile." Now, combine the root for "bile" with the one for "bladder," which is *cyst* (**SIST**). Finally, we build a new term, *cholecyst* (**KOH**-luh-sist). The cholecyst is literally the "bile bladder" or the "gall bladder."

Figure 12.3 reveals the close working partnership between the liver and the cholecyst (gall bladder). Thousands of *hepatocytes* (**HEP**-ah-toh-**sights**) or "liver" (*hepat*) "cells" (*cytes*) constantly synthesize and secrete bile. The bile constantly drips down into the *right and left hepatic* (heh-**PAT**-ik) *ducts*, which also carry many of the other secretions from the liver.

The bile takes a curve off from the hepatic ducts, and moves into the *cystic* (**SIST**-ik) *duct*. This duct is named for its physiology of supplying a *cyst* or "bladder" (specifically, the *cholecyst* or gall "bladder"). The cystic duct then carries the secreted bile into the cholecyst. The cholecyst (gall bladder) is a muscular-walled pouch that stores the bile coming down to it from the liver overhead.

Organ 1

We can now state *the wonderful Liver–Gall Bladder "Bile Hand-holding" Rule*: "Bile is continuously **produced** and **secreted** by the **liver**, but is then temporarily **stored** and **released** into the small intestine by the ***cholecyst (gall bladder)***."

THE DUODENUM: FINAL DESTINATION FOR BILE AND PANCREATIC JUICE

Note from Figure 12.3 that bile leaves the cholecyst (gall bladder) storage sac, moves out through the cystic duct, and then into the common bile duct. The common bile duct carries the bile down to its final destination – the *duodenum* (dew-**AH**-den-um). The duodenum is the first or most proximal (closest to the stomach) portion of the small intestine.

The duodenum is named for an observation made by some Ancient Latin dissectors: it is about "twelve" (*duoden*) finger-widths (finger-breadths) long. For such a relatively small area, as you can plainly see from the figure, the duodenum is very *busy*, isn't it! After all, not only does it receive bile from the liver and gall bladder (cholecyst), but it also gets the *pancreatic* (pan-kree-**AT**-ik) *juice*.

The pancreatic juice is the main secretion of the *exocrine* portion of the pancreas, which releases it into the *pancreatic duct*. (You may recollect from Chapter 8 that the endocrine or "internally secreting" portion of the pancreas produces the hormones, insulin and glucagon.)

Both the bile flowing through the common bile duct, as well as the pancreatic juice moving through the pancreatic duct, join together at the *hepatopancreatic* (heh-**PAT**-oh-pan-kree-**AT**-ik) *sphincter* (**SFINGK**-ter). (A sphincter, in general, is a ring of muscle tissue that strongly closes off the lumen of some tube or passageway.) When the hepatopancreatic sphincter relaxes, bile and pancreatic juice flow through the *duodenal* (dew-**AH**-deh-nal) *papilla* (pah-**PIL**-lah). The duodenal papilla is a "little nipple or pimple" (*papill*)-like projection with a hole in its center. Consequently, both bile and pancreatic juice drip into the duodenum through this hole in the duodenal papilla.

BILE SQUIRTS OUT OF THE GALL BLADDER: IT'S ALL A MATTER OF *MOTILITY*!

Unfortunately, Figure 12.3 only showed us *where* the bile was being carried, after it was secreted (by the liver) and temporarily stored (by the gall bladder). But it did not reveal exactly *why* the stored bile was suddenly released from the cholecyst in the first place. For this we need to discuss a particular physiological control mechanism, such as negative feedback control (Chapter 2) over hormone secretion.

And this negative feedback control mechanism or system needs to focus upon still another general function of the digestive tube – its *motility* (moh-**TIL**-ih-tee) or "ability" (-*ity*) to "move" (*motil*). "You mean, if I'm having a bowel *movement*, then my intestines are carrying out the general function of *motility*?" Good insight, Baby Heinie!

In the case of the cholecyst or gall bladder, the motility (ability to move) lies with the capacity of its muscular wall to contract. When the muscular wall contracts, it squeezes or compresses the gall bladder lumen. This squeezing action squirts a slug of stored bile out of the cholecyst, and eventually down into the duodenum. [**Study suggestion:** To visualize this effect, think of a rubber bulb attached to one end of a plastic tube, as found in the kitchen, say, on a turkey or poultry baster. When you initially squeeze the bulb, the negative pressure (suction) draws a quantity of gravy or broth up into the bulb (gall bladder), where it is temporarily stored. The gravy will stay inside the bulb until your hand gets "excited" and squeezes the filled bulb. This squeezing action squirts a slug of gravy (stored bile) out of the bulb (cholecyst) and into the plastic tube (common bile duct).]

STIMULATING THE GALL BLADDER TO CONTRACT: THE DUODENUM SECRETES A HORMONE

"But *what* is it, exactly, that stimulates or excites the hand (smooth muscle tissue in gall bladder wall) to squeeze the bulb (contract the gall bladder)?" Good question, Baby Heinie! Take a look at Figure 12.4(A).

Soon after you eat a heavy meal, especially one containing a lot of fat, a large amount of fatty chyme leaves the stomach and enters the duodenum. The original stimulus for gall bladder contraction, then, is the presence of a large amount of fatty chyme within the duodenum. The fatty chyme stretches the duodenal wall and stimulates a population of endocrine gland cells located there. This group of gland cells secretes a hormone called *cholecystokinin* (**KOH**-luh-sist-oh-**KIGH**-nin) or *CCK* for short.

The word cholecystokinin literally means "a substance" (-*in*) that "moves" (*kin*) the "gall bladder" (*cholecyst*). Specifically, cholecystokinin (CCK) enters the bloodstream after being secreted by the duodenal gland cells. The bloodstream carries the CCK up to the gall bladder. True to its name, the CCK stimulates the smooth muscle in the cholecyst wall to contract and "move" (*kin*). Thus, cholecystokinin, because of its kinetic ("pertaining to

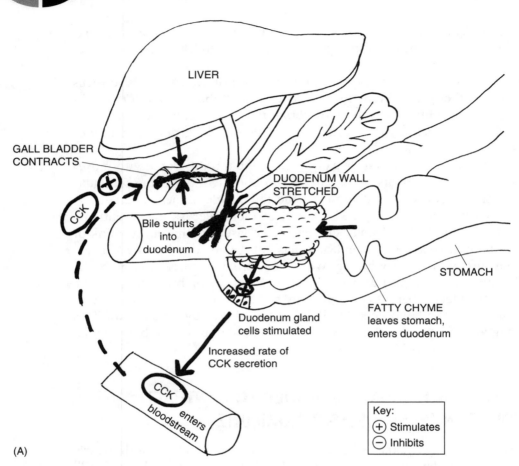

(A)

Fig. 12.4   Negative feedback system for CCK secretion and contraction of gall bladder. (A) Fatty chyme triggers CCK (cholecystokinin) release, which stimulates gall bladder to squirt out stored bile. (B) Emulsification of fat inhibits duodenal gland cells. Rate of CCK secretion greatly decreases. Gall bladder relaxes.

movement'') effect, triggers the motility of the gall bladder wall. As a result of this squeezing action, a slug of stored bile is squirted out of the gall bladder, down through the common bile duct, and into the duodenum.

The bile, as we have mentioned, is a *detergent* substance (somewhat like the soapy detergents used to clean the grease from our dirty clothes and dishes). A chemical detergent (such as bile) does its job by causing *emulsification* (ih-**mul**-suh-fuh-**KAY**-shun). Emulsification is a "milking out" of fat from the rest of the chyme in the duodenum. The big globs of fat are then broken down into a foam of tiny fat globules. (Consult Figure 12.4, B.)

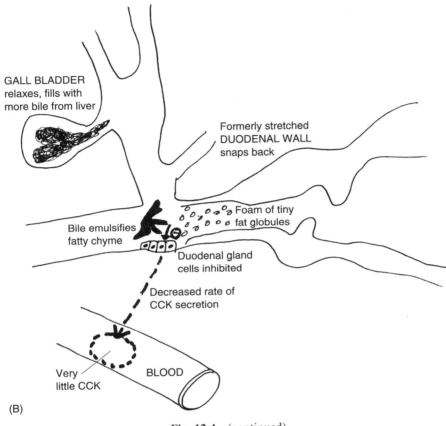

GALL BLADDER
relaxes, fills with
more bile from liver

Formerly stretched
DUODENAL WALL
snaps back

Foam of tiny
fat globules

Bile emulsifies
fatty chyme

Duodenal gland
cells inhibited

Decreased rate of
CCK secretion

Very
little CCK

BLOOD

(B)

Fig. 12.4   (continued)

Once the big globs of fat in the chyme are broken down, the duodenal wall snaps back from its formerly stretched, fat-swollen condition. This creates a negative feedback effect upon the duodenal gland cells. Since the original stimulus of too much fat and stretching of the duodenum has been corrected or removed, further secretion of CCK from the duodenal cells is inhibited.

Too Much Acid? The Duodenum Just Keeps "Secretin'"!

Being situated just downstream from the stomach, having to deal with fatty chyme is only *one* of the digestive challenges being faced by the duodenum!

As you may recall from our earlier discussions of body acids and bases (Chapter 3), the walls of the stomach secrete HCl (hydrochloric acid). The HCl rapidly dissociates (breaks down) into H^+ and Cl^- ions. The hydrogen (H^+) ions are extremely reactive, allowing them to chemically digest many of the organic molecules found within the stomach chyme.

"Why doesn't the stomach digest *itself*, Professor? It has all those millions of highly reactive H^+ ions moving around, you know!" The *mucosa* (myoo-**KOH**-sah) – mucous membrane lining – of the stomach does effectively protect itself from the damaging effects of HCl by secreting a 1-millimeter-thick *alkaline mucous film*. This basic, high-pH film neutralizes many of the potentially dangerous H^+ ions hitting the stomach wall.

THE DUODENUM PROTECTS ITSELF

Organ 1

The poor duodenum, however, has *no* protective mucous film on its lining! Hence, *duodenal ulcers* are more common than *gastric* (**GAS**-trik) or "stomach" (*gastr*) *ulcers*. The chyme coming into the duodenum is usually an acidic or acid-bearing chyme, carrying with it millions of H^+ ions from the stomach HCl. These highly reactive H^+ ions can wear away the duodenal mucosa and create ulcers (open sores). Such duodenal ulcers can be very painful, and even dangerous.

But the duodenum is fairly well prepared with its own hormonal defense against excess gastric acid. Figure 12.5 shows another negative feedback control mechanism, this one involving a stimulus of a low pH (increased acidity or $[H^+]$) of the chyme present within the duodenum.

The presence of acid chyme stimulates another group of duodenal gland cells (different from those that produce CCK). This population of gland cells responds to the acid by greatly increasing their release of *secretin* (see-**KREE**-tin). As its name suggests, secretin was the very first hormone to be discovered. Hence the word secretin literally means "a substance" (-*in*) that is "secreted" (*secret*) into the bloodstream. (View Figure 12.5, A.)

Secretin molecules are then circulated out to the liver, the pancreas, and to the stomach. Secretin inhibits further HCl secretion by the *gastric glands*. Hence, the rate at which additional harmful acid moves from the stomach into the duodenum is slowed. (See Figure 12.5, B.)

Both the hepatocytes in the liver and the acinar cells in the pancreas respond to secretin by releasing more *sodium bicarbonate* ($NaHCO_3$) into the common bile duct. (Study Figure 12.6.) Recall (Chapter 3) that sodium bicarbonate is a weak base or alkali that accepts H^+ ions. [**Study suggestion:** Before reading any farther, why not flip back to Chapter 3 and refresh your

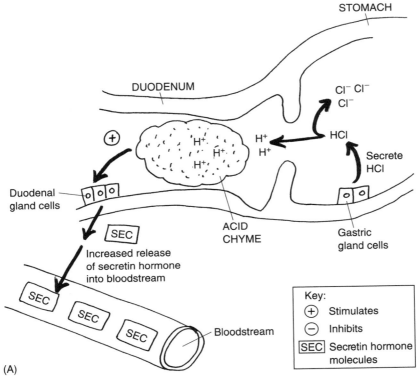

STOMACH

DUODENUM

Cl^- Cl^-
Cl^-

H^+
H^+

HCl

H^+
H^+
H^+
H^+
H^+

Secrete
HCl

(+)

Duodenal
gland cells

SEC

Increased release
of secretin hormone
into bloodstream

ACID
CHYME

Gastric
gland cells

SEC

SEC

SEC

Bloodstream

Key:
(+) Stimulates
(−) Inhibits
SEC Secretin hormone
molecules

(A)

Fig. 12.5   The duodenum protects itself from acid by releasing more secretin. (A) Acid chyme
in duodenum stimulates release of secretin hormone into bloodstream.

memory on acids, bases or alkali, acid–base homeostasis, and pH? *Then* come
back right here!]

By stimulating the liver and pancreas to secrete more sodium bicarbonate
($NaHCO_3$) into the duodenum, secretin is providing a weak base that can
react with the excess stomach acid present in its lumen. And, as shown in
Figure 12.6, this weak base allows the *intestinal neutralization equation* to take
place. By this reaction, the very strong, corrosive acid – HCl – is essentially
replaced by a weak, relatively non-corrosive acid – H_2CO_3 or carbonic acid.

"Oh, I remember carbonic acid from Chapter 11 and gas transport,
Professor! I remember that carbonic acid was formed as the result of CO_2
reacting with H_2O inside the red blood cells!" That's correct, Baby Heinie.
"So, H_2CO_3 must be pretty harmless, because it's produced every minute
inside our RBCs!"

As the released sodium bicarbonate neutralizes a lot of the strong HCl, the
pH of the duodenum rises, and its level of harmful acidity falls. This final

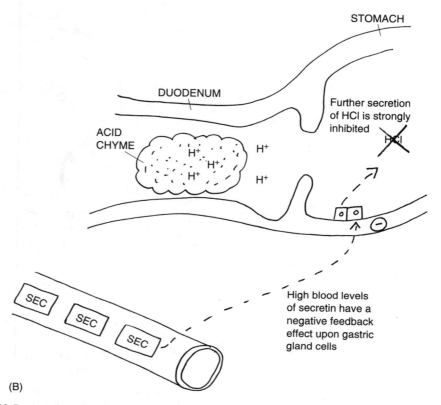

(B)

Fig. 12.5 (continued) (B) Secretin has a negative feedback effect upon gastric gland cells, thereby inhibiting their secretion of HCl (hydrochloric acid).

response removes the conditions of the initial stimulus, so that the endocrine gland cells in the duodenal wall are inhibited from releasing any more large amounts of secretin.

More Digestive Enzymes: The CCK "Bonus" Effect

We have now discussed two digestive hormones, cholecystokinin (CCK) and secretin. Remember that CCK stimulates contraction of the gall bladder (cholecyst) to release bile. And recall that secretin stimulates the liver and pancreas to secrete more sodium bicarbonate ($NaHCO_3$) base into the duodenum, thereby neutralizing excess acid.

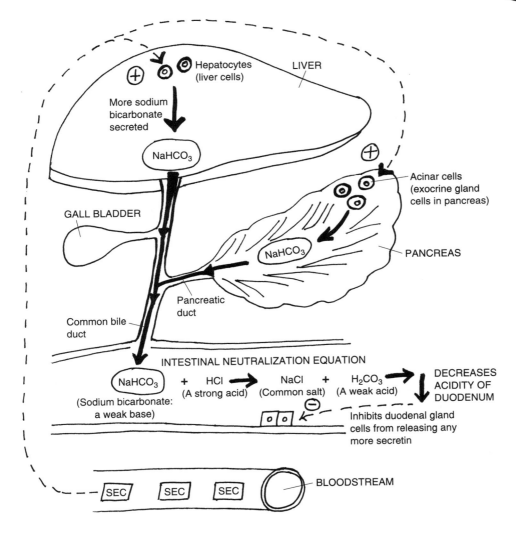

Fig. 12.6 Intestinal neutralization: Secretin stimulates hepatocytes (in liver) and acinar cells (in pancreas) to increase their secretion of sodium bicarbonate (NaHCO₃) into the duodenum. Excess stomach HCl is neutralized, and duodenum becomes less acidic.

There is, however, still another effect of cholecystokinin that we have not mentioned. For the sake of convenience, let's call it *the CCK "Bonus" Effect upon the Pancreas: In addition to stimulating the gall bladder, cholecystokinin provides a "bonus" (additional) action upon the pancreas. CCK stimulates the pancreas to increase its secretion of **digestive enzymes** into the duodenum.*

Organ 2

Now, the duodenum is where it all seems to come together! It is here that the chemical digestion of carbohydrates is finished. *Amylase* (**AM**-ih-**lace**) enzymes or "starch" (*amyl*)-"splitters" are secreted in greater amounts by the pancreas (due to CCK stimulation). These starch-splitting enzymes are carried into the duodenum within the pancreatic juice. Here they complete the catabolism (breakdown) of starches into simple sugars, such as glucose, which are then absorbed into the bloodstream.

Similarly, CCK stimulates the pancreas to release more *proteases* (**PROH**-tee-**ay**-sez) or "protein-splitters" into the duodenum. Here the proteases complete the chemical digestion of protein fragments into individual amino acids, which are absorbed into the blood.

Finally, CCK stimulates the pancreas to secrete more *lipases* (**LIE**-pay-sez) or "fat-splitters" into the duodenum. Consequently, the chemical digestion of fats into individual fatty acid and glycerol molecules is accomplished here. These final products of fat digestion are likewise absorbed.

Villi and Microvilli in the Small Intestine Wall

"Wow, Professor! I'm really *impressed*! When you said that the pancreas does it *all*, you were right on!" A look at Figure 12.7, which provides a detailed view of the small intestine wall, should impress you even more, Baby Heinie.

We have already mentioned one major *tunic* (**TOO**-nik) or "coat" in the digestive tube wall – the mucosa. We said that it was a mucous membrane. Figure 12.7 provides a detailed close-up of the mucosa in the small intestine wall. Here we see two special modifications of the mucosa. These are called the *villi* (**VIL**-ee) or "tufts of hair," and the *microvilli* (**my**-kroh-**VIL**-ee) or "tiny tufts of hair." We can see from the close-up that each single *villus* (**VIL**-us) does somewhat resemble a chubby "tuft of hair" projecting into the lumen of the tube from the mucosa.

And literally "below" (*sub*-) the "mucosa" is a second layer in the wall, called the *submucosa* (**SUB**-myew-koh-sah). This layer is highly vascular and supplies numerous blood vessels, into which nutrients enter after they are absorbed across the surface of the mucosa.

Because each villus has such a large surface area available for absorbing digested nutrients, the thousands of villi within the small intestine wall allow it to be highly efficient in its absorption. Note also that the surface of each villus is studded with dozens of microvilli. These microscopic, hair-like projections increase the available surface area for absorbing nutrients even more!

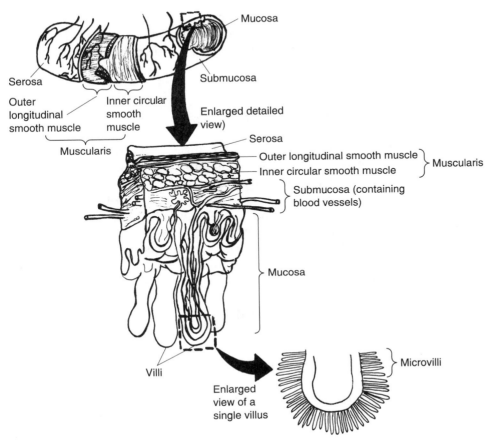

Mucosa

Serosa

Outer
longitudinal
smooth muscle

Inner circular
smooth
muscle

Submucosa

Enlarged detailed
view)

Muscularis

Serosa

Outer longitudinal smooth muscle ⎫
Inner circular smooth muscle   ⎬ Muscularis

Submucosa (containing
blood vessels)

Mucosa

Villi

Enlarged
view of a
single villus

Microvilli

Fig. 12.7   Major tunics in the small intestine wall.

GI Tube Movements

Observe in Figure 12.7 that there are two outer, more peripheral tunics in the digestive tube wall, in addition to the mucosa and submucosa. These are called the *muscularis* (**mus**-kyoo-**LAIR**-us) and the *serosa* (**see-ROH**-sah). The muscularis is the smooth muscle tissue area in the digestive tube wall. The serosa, in contrast, is a *serous* (**SEER**-us) or "watery" membrane. While the serosa is most important for moistening the outer surface of the GI tube wall, the muscularis is critical for its motility (ability to move and constrict the tube).

Figure 12.7 reveals that there are two thinner layers within the muscularis. These are the outer *longitudinal smooth muscle layer*, and the inner *circular smooth muscle layer.* The longitudinal smooth muscle layer contains muscle fibers running "lengthwise" (longitudinally) along the GI tube wall, while the circular smooth muscle layer has its fibers arranged in circles around the tube lumen.

SMOOTH MUSCLE PACEMAKER CELLS

Smooth muscle tissue in the walls of the digestive tube, like the cardiac muscle tissue in the wall of the heart (Chapter 9), contains pacemaker cells that automatically depolarize or excite themselves at a certain rate or rhythm. These *smooth muscle pacemaker cells* are modified smooth muscle fibers. They are generally located in the outer longitudinal smooth muscle layer of the muscularis. The smooth muscle pacemaker cells automatically depolarize and become leaky to Na^+ ions at a certain rate. This spontaneous depolarization of the pacemaker cells keeps summating (adding together) until the threshold is reached for firing *smooth muscle action potentials* – traveling waves of electrochemical excitation.

After they are sufficiently stimulated by smooth muscle action potentials from the outer longitudinal smooth muscle layer, the inner circular smooth muscle contracts and narrows (constricts) the digestive tube lumen. This constriction or narrowing serves to push the food, chyme, or feces through the digestive tract.

Remember this: *In general, the outer* **longitudinal** *smooth muscle tissue layer contains* **pacemaker cells** *that create* **smooth muscle action potentials**, *thereby* **exciting** *the inner circular smooth muscle tissue layer. Once excited, the inner* **circular** *smooth muscle layer* **contracts and constricts** *the digestive tube lumen (open interior). Finally, this constriction of the lumen pushes material through the tube, toward the anus.*

Organ 3

SPECIFIC TYPES OF DIGESTIVE TUBE MOVEMENTS

Now we will consider some specific types of movements of the digestive tube. Figure 12.8 will provide us with a toothpaste tube model of the digestive tract to help illustrate each type.

Peristalsis

The most basic type of movement of the digestive tube is *peristalsis* (per-ih-**STAL**-sis). Peristalsis is literally a "contraction around" a particular point of the digestive tube, such that the contained material is pushed onward through the lumen. We take our big tube of toothpaste, turn it upside down, and remove the cap. At first, all points along the tube are relaxed (Figure 12.8, A). Place both hands around the bottom of the tube so that they encircle it (like the circular smooth muscle layer). We (the smooth muscle pacemaker cells) count, "One ... two ... three ... constrict!" As we squeeze the tube at a single point (peristalsis), the toothpaste gets pushed closer toward the cap end (see Figure 12.8, B.)

Peristalsis begins in the lower 2/3 of the *esophagus* (eh-**SAHF**-uh-**gus**) or "gullet," the muscular tube leading into the stomach. It continues through the stomach.

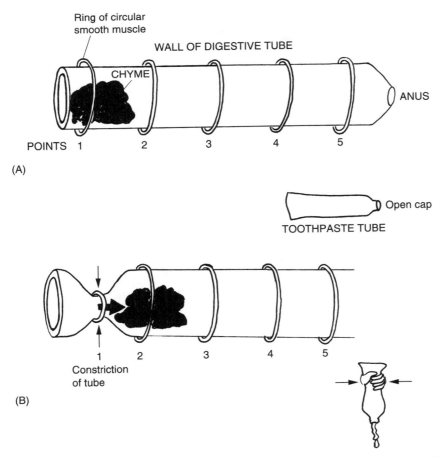

Fig. 12.8 GI tract motility and the toothpaste tube model. (A) All points along tube are relaxed (not constricted). (B) Peristalsis occurs at point 1.

Segmentation

An alternative type of digestive tube movement is called *segmentation* (**seg-men-TAY**-shun). Segmentation is the simultaneous contraction and narrowing of several non-adjacent (non-neighboring) portions of the digestive tube, such that *segments* of tube are made between them (see Figure 12.8, C).

Segmentation mainly occurs in the small intestine, where it mixes the chyme with digestive enzymes as it is churned back and forth between tube segments.

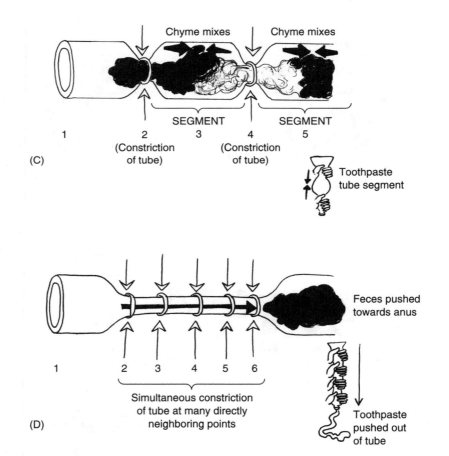

Fig. 12.8   (continued). (C) Segmentation: simultaneous constriction at points 2 and 4 creates segments of small intestine, thoroughly mixing the chyme within each segment. (D) Mass movement (mass peristalsis): simultaneous constriction at points 2 through 6 pushes feces onward.

Mass movement (mass peristalsis)

The third type of digestive tube movement is called *mass movement or mass peristalsis*. This is the simultaneous contraction and narrowing of several directly adjacent (side-by-side) portions of the digestive tube, such that the feces are pushed in a large *mass* from one section of the colon to another (see Figure 12.8, D).

Mass movement (mass peristalsis) occurs in the colon (large intestine) only three or four times a day, usually during or soon after eating. It is the main process whereby our heavy "dregs" (feces) are moved along, getting us ready to defecate them.

THE DEFECATION REFLEX

After mass movements powerfully propel some feces all the way through the large intestine, they drop down through the *sigmoid* (**SIG**-moyd) *colon*, and into the *rectum* (**REK**-tum). The rectum is a "straight" (*rect*) muscular-walled tube that leads into the anus.

Two sphincters within the rectum stand between the stools and the anus. A higher *internal anal sphincter* is composed of smooth muscle, so it is not under our voluntary control. Conversely, the lower *external anal sphincter* is made up of striated (cross-striped) muscle tissue. Hence, its contraction and relaxation *is* under our conscious control.

The wall of the upper rectum is stretched as feces enter it, thereby triggering the *defecation reflex*. This involves the automatic relaxation of the internal anal sphincter when the rectal wall has stretched out far enough with its load of feces. The feces drop down to hit the external anal sphincter, also automatically.

After the defecation reflex, it is up to the conscious will of the individual to decide when and where to relax the external anal sphincter, and thus to defecate.

Vomiting and Diarrhea: Digestive Movements That Can Upset Acid–Base Balance

Our earlier discussions have informed us about the role of the duodenum and of $NaHCO_3$ base from the pancreas, in neutralizing HCl from the stomach. A follow-up question that needs to be asked is, "What if there is an excessive amount of motility in the digestive tube, either at its upper or lower end? Will this have any effect upon body acid-base balance?"

VOMITING AND METABOLIC ALKALOSIS

When a person vomits, a large quantity of highly acidic chyme is thrown up from the stomach. Hence, a big volume of HCl will tend to be lost. But most of the $NaHCO_3$ base in the duodenum will remain. The intestinal neutralization process can no longer work properly, and *metabolic alkalosis* may result.

Organism 1

Metabolic alkalosis is an abnormal condition of too much body base or alkali, or not enough acid, due to *metabolic*, that is, non-respiratory, causes. One of the most important non-respiratory causes of metabolic alkalosis is vomiting.

DIARRHEA AND METABOLIC ACIDOSIS

When a person has diarrhea, they excrete a large amount of watery stool from their intestines. Because sodium bicarbonate ($NaHCO_3$) is present in the small intestine, much of it will be lost with the watery feces. Yet the HCl in the stomach will mainly be left behind. This can produce *metabolic acidosis*.

Organism 2

Metabolic acidosis is an abnormal condition of too much body acid, or not enough base or alkali, due to *metabolic*, that is, non-respiratory, causes. One of the chief non-respiratory causes of metabolic acidosis is diarrhea.

Quiz

Refer to the text in this chapter if necessary. A good score is at least 8 correct answers out of these 10 questions. The answers are listed in the back of the book.

1. The muscular ring through which one defecates:
   (a) Hyoid
   (b) Anus
   (c) Diaphragm
   (d) Sigmoid colon

2. Defecation is alternately known as:
   (a) Egestion
   (b) Digestion

    (c) Absorption
    (d) Indigestion

3. Feces are like "dregs" because they are:
    (a) Brown like mud or sediment
    (b) Frequently passed at the same time as one urinates
    (c) Held aloft by Roman scholars
    (d) A real drag to be around!

4. Dietary fiber is not absorbed, since:
    (a) People really don't like to eat it
    (b) Gastric enzymes chop it up into fragments that are too tiny
    (c) It often contains cellulose, which is largely indigestible by humans
    (d) There is too much water stuck to its surface

5. If a large cholelith blocks the common bile duct, then the feces will:
    (a) Probably be purple and have a low fat content
    (b) Appear slate-gray and have fatty streaks in them
    (c) Likely not be affected
    (d) Always be passed as watery diarrhea

6. Bile is actually secreted by the:
    (a) Cholecyst
    (b) Duodenum
    (c) Pancreas
    (d) Liver

7. Cholecystokinin's main action:
    (a) Release of $NaHCO_3$ from the pancreas
    (b) Powerful stimulation of mass movements
    (c) Terrifying constipation
    (d) Release of bile stored in the gall bladder

8. Secretin's primary action:
    (a) Causes the hypothalamus to secrete RHs
    (b) Removal of bile pigments from the bloodstream
    (c) Stimulates secretion of sodium bicarbonate by liver and pancreas
    (d) Activation of the defecation reflex

9. Microvilli have as their important characteristic:
    (a) Vastly increase surface area for absorption
    (b) Cut the wasting of ATP for unneeded cell repairs
    (c) Greater motility of the GI tract
    (d) Boost the rate of smooth muscle pacemaker cells

10. Helps efficiently mix chyme with digestive enzymes in the small intestine:
    (a) Mass peristalsis
    (b) Vomiting
    (c) Segmentation
    (d) Peristalsis

Body-Level Grids for Chapter 12

Anatomy and *Biological Order* Fact Grids for Chapter 12:

A

ORGAN
Level

1

ORGAN SYSTEM
Level

1	2

Physiology and *Biological Order* **Fact Grids for Chapter 12:**

ORGAN
Level

1	2
3	

ORGAN SYSTEM
Level

1

Physiology and *Biological Disorder* Fact Grids for Chapter 12:

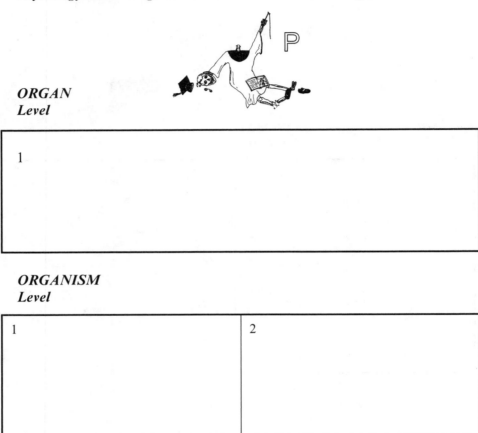

ORGAN
Level

1

ORGANISM
Level

1	2

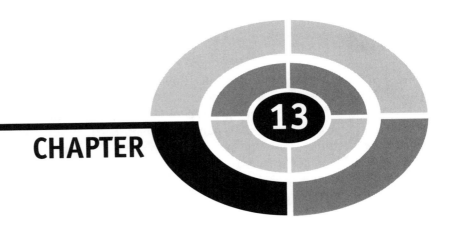

The Genitourinary System: It Allows Us to Release Fluids and "Beget" New Life!

It is fitting that we bring *PHYSIOLOGY DEMYSTIFIED: A Living Pathway Through Bodyspace* to a close with an attempt to explain how we "beget" life. To do so will require us to discuss the combined *genitourinary system*, which can also be called the *urogenital* (**you**-roh-**JEN**-ih-tal) *system*.

"What's that?"

We mean by "beget" the exact translation of the Latin word root *genit*, which means "beget or produce" new offspring. Thus, we are going to be considering the physiology of the human *genital* (**JEN**-ih-tal) *organs*, which allow us to carry out this critical survival function of reproduction. Of course, the genital system is often referred to as the *reproductive system*.

But in covering the genital organs, we will also be obliged to study the *urinary* (**YOUR**-ih-**nair**-ee) *system*, because some of the organs in these two different systems are *shared*. In the male *penis* (**PEA**-nis) or "tail," especially, this fact is true. It both delivers the "seed animals" called *spermatozoa* (sper-**mat**-uh-**ZOH**-ah), as well as excreting the *urine* from the body. So the penis is neither just a genital organ nor just a urinary organ. Rather, it is a *combined* genitourinary organ.

Our approach is therefore summarized *anatomically* as:

A

THE GENITOURINARY	=	GENITAL	+ URINARY
SYSTEM (UROGENITAL		(REPRODUCTIVE)	SYSTEM
SYSTEM)		SYSTEM	

Organ System 1   For a detailed discussion of these aspects of G-U body structure, please consult our companion volume, *ANATOMY DEMYSTIFIED*.

The Notion of Physiological Pathways Within the Great Body Pyramid

As you may recollect (Chapter 1), Bodyspace was basically equated to the existence of a highly orderly *internal environment* within the human organism. We modeled Bodyspace geometrically, as a series of stacked grids creating a Great Body Pyramid. And *physiology* was viewed as a curving "Pathway of Life" that passes through these stacked grids in the human Bodyspace Pyramid.

HOMEOSTASIS AND NEGATIVE FEEDBACK AS CURVING PATHWAYS

Two of the core concepts in human physiology – *homeostasis* and *negative feedback systems or cycles* – can also be interpreted as curving pathways that proceed through Bodyspace for certain time periods:

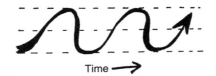

Time →

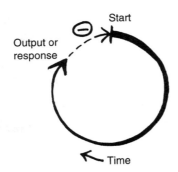

The Three Genitourinary Pathways

Pathways are fundamentally *physiological* concepts, since they imply that a body *function* is occurring – something is *moving* through the pathway. In our present study, we can talk about three *genitourinary pathways*. It will also be of great interest to examine *what* is moving through these G-U pathways, as well as *how* and *why* they are moving through them.

1. THE URINARY PATHWAY AND ITS EXIT FROM THE PYRAMID

The *urinary pathway* is simply the pathway followed by the *urine* as it is being excreted from the body. In an abstract sense (Figure 13.1), the urinary pathway can be drawn as a curving line that *leaves* the Great Body Pyramid at the Chemical Level of organization (molecules, atoms, and subatomic particles). Urine is technically defined as an *amber* ("yellowish-brown"), salty, slightly acidic, watery fluid excreted by the kidneys.

The actual word, *urine*, however, can also translate from ancient Greek to mean "presence of *urea*" (you-**REE**-ah). Urea is the main nitrogen-containing solute found within the urine. It is a white crystal molecule that is formed

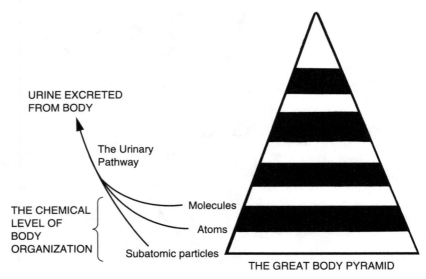

URINE EXCRETED
FROM BODY

The Urinary
Pathway

THE CHEMICAL
LEVEL OF
BODY
ORGANIZATION

Molecules

Atoms

Subatomic particles

THE GREAT BODY PYRAMID

Fig. 13.1   The urinary pathway exiting at the Chemical Level of the Great Body Pyramid.

from amino acids catabolized by our cells for ATP energy. Thus, urea is the final breakdown product of protein metabolism in the body.

Organ 1

Deadly uremia

Urea is also a highly toxic *metabolic poison*. When *renal* (**REE**-nal) or "kidney" *failure* occurs for any reason, the urea constantly being produced by protein metabolism is not sufficiently excreted from the body in the urine. The urea concentration thus builds up to excessive levels within the bloodstream. This renal pathophysiology soon results in *uremia* (you-**REE**-me-ah), also called *uremic* (you-**REE**-mik) *poisoning*.

Symptoms of uremia (uremic poisoning) include nausea, vomiting, headache, dizziness, and dimness of vision. If allowed to progress untreated, uremia will eventually kill the patient due to *uremic coma and convulsions.*

Urine composition

Urine consists of about 95% water and 5% dissolved solutes. Besides urea, some of the other main solutes in the urine are sodium chloride (NaCl) – explaining its saltiness – and lots of hydrogen (H^+) ions – explaining its slight acidity. The amber color of urine is mainly due to a certain molecule related to the breakdown of bilirubin from the bile (Chapter 12). It is called *urobilin* (**you**-roh-**BY**-lin), a brown pigment that colors both the stools and the urine.

Urinary pathway summary

In conclusion, we can say that the urinary pathway is largely a "getting rid of urea" pathway. And as a by-product of ridding the body of poisonous urea, the urine also serves as a pathway for excreting other substances that are worthless or potentially harmful.

URINARY PATHWAY	= "A Getting Rid of *Urea* Pathway"	= An *Excretory* Pathway Leaving the Body Pyramid at the *Chemical* Level (*Molecules*: H_2O, Urea; *Atoms* (*Ions*): Na^+, Cl^-, H^+)

Molecule 1

2. THE SPERM PATHWAY AND ITS EXIT FROM THE PYRAMID

The male genital or reproductive pathway is rather *vaguely* named, isn't it? It would be much more helpful and information-rich to specify exactly *what* the male genital/reproductive pathway is *carrying*.

The *spermatozoa* (sper-**mat**-uh-**ZOH**-ah) or sperm cells (literally called "seed animals" in Latin) are the things being carried! The intriguing explanation for this very strange name becomes much more clear when we examine the microanatomy of a single *spermatozoon* (sper-**mat**-uh-**ZOH**-un) or sperm cell under a microscope.

Figure 13.2 (A) reveals that a spermatozoon does, indeed, somewhat resemble a tadpole or some other little swimming "animal"! Its body consists of three major sections: a pointed, torpedo-shaped *head*, a *neck or midpiece*, and a long, whip-like *tail*.

The head consists of the cell nucleus, plus a stocking cap-shaped covering over the front called the *acrosome* (**AK**-ruh-sohm) or "end-body." The acrosome contains special digestive enzymes that allow the head of the sperm to dissolve its way into an *ovum* (**OH**-vum) or "egg" (*ov*) cell of the female during fertilization.

The neck or midpiece consists of a group of tightly coiled mitochondria, which produce the ATP energy required for whipping the tail. The tail is basically a *flagellum* (flah-**JELL**-um) or "whip"-like strand that propels the spermatozoon forward.

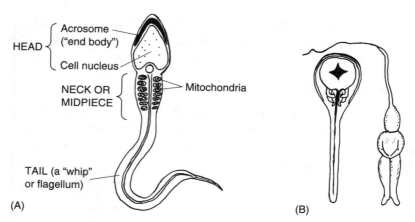

HEAD { Acrosome ("end body")
Cell nucleus

NECK OR MIDPIECE — Mitochondria

TAIL (a "whip" or flagellum)

(A)   (B)

Fig. 13.2   Differing views of a spermatozoon: old versus new. (A) The basic anatomy of a spermatozoon. (B) Early ideas about the sperm as a homunculus ("little man").

Physiology of the spermatozoa

The primary function of spermatozoa, of course, is fertilization of a mature ovum in a female. When fertilization occurs, the 23 chromosomes in the nucleus of the sperm-head join with the 23 chromosomes in the nucleus of the ovum.

Before modern times, however, this fertilizing function of the spermatozoa was not at all so obvious. Consider, for example, the bizarre illustrations in Figure 13.2 (B). These show two different versions of a spermatozoon drawn in the very early days of the light microscope. According to the *Preformation Theory* of the 1700s and 1800s, the human *embryo* (**EM**-bree-oh) was "preformed" (already formed) in a very miniature body, within either a spermatozoon or an ovum, *before* fertilization even took place! Thus, the drawings show a tiny *homunculus* (hoh-**MUN**-yoo-lus) or "little man" encased within each sperm cell. [**Study suggestion:** Review the modern description of the sperm cell, and try to guess why one of the *homunculi* (hoh-**MUN**-kyoo-lie) in the figure seems to have a big head with a star-like region in it. What do you think this central, star-like region *actually* represents in the sperm cell?]

Addition of the semen

Spermatogenesis (sper-mat-uh-**JEN**-eh-sis) is the "production of" (*-genesis*) mature "sperm" (*spermat*) cells. Spermatogenesis takes place within the highly coiled, tiny tubes of the *testes* (**TES**-teez). These tiny tubes are called the *seminiferous* (sem-ih-**NIF**-er-us) *tubules* (**TOO**-byools).

A healthy young male in his 20s may produce up to 400 million spermatozoa every day! They are *ejaculated* (ih-**JACK**-yuh-**lay**-ted) or "thrown out"
of the *urethra* (yew-**REETH**-rah) at the tip of the penis. When the swimming
of the spermatozoa is not assisted by ejaculation, they move at a maximum
speed of about 3 millimeters every hour. These sperm do not have very long
to reach an ovum and attempt to fertilize it! Ejaculated spermatozoa have a
life expectancy of just 48 to 72 hours.

The spermatozoa, however, are not the only things present within the
ejaculate (ee-**JACK**-yoo-lut). After all, they do have to swim *through* something *wet* – not just thrash their tails around within hollow tubes filled with
air! What the sperm are swimming through is the *seminal* (**SEM**-ih-nal) *fluid*
or *semen* (**SEE**-mun). The semen (seminal fluid) is the fluid portion of the
ejaculate. Semen is the thick, milky, sugar-rich, basic pH fluid which sus-
pends and provides nutrients for the "seeds" (living sperm cells).

Semen is the fluid added to the sperm when they are being ejaculated. Most
of the fluid is secreted by two types of *accessory* ("added-on") *reproducive
organs*: the pair of *seminal vesicles* ("seed bladders") and the single *prostate*
(**PRAH**-state) *gland*.

The male ejaculate is about 2 to 5 milliliters in volume. Approximately
90% of this ejaculate is semen (seminal fluid), with the other 10% being
spermatozoa. The semen or seminal fluid, in turn, consists of about 60%
seminal vesicle fluid and 30% *prostate gland fluid*.

The *sperm count* has a lower normal limit of 50 million spermatozoa/ml
of ejaculate, and an upper normal limit of 120 million spermatozoa/ml of
ejaculate.

Semen summary

To sum things up about the spermatozoa and semen, we state the following
regarding the ejaculate:

| THE EJACULATE (2–5 ml in volume) | = | SEMEN OR *SEMINAL* FLUID (60% *seminal* vesicle fluid; plus 30% prostate gland fluid; 10% from other sources) | + | SPERMATOZOA (Ejaculated sperm cells) |

A

Tissue 1

The sperm pathway toward the ovum

The *sperm pathway leading to fertilization* is the pathway followed by the
spermatozoa within the semen after they have been ejaculated. When we
refer back to the model of the Great Body Pyramid, the male *ejaculate* can

be considered an example of the *tissue* level of biological organization. [**Study suggestion:** Why do you think that the male ejaculate best represents the tissue level of organization within the Pyramid? What is the basic definition of any tissue, and why do the characteristics of the male ejaculate fit this definition pretty well?]

In an abstract sense (Figure 13.3), the sperm pathway leading to fertilization can be pictured as a pathway of secretion of ejaculate (semen + spermatozoa) leading out of the tissue level of the Great *Male* Body Pyramid. If the sperm pathway is aimed at fertilization, then the secreted ejaculate travels over to the *organ level* of a nearby Great *Female* Body Pyramid. The ejaculate goes into the organ level, because it enters the female *vagina* (vah-**JEYE**-nah). The vagina is a reproductive *organ* of the female, which literally serves as a "sheath" for an inserted penis during sexual intercourse.

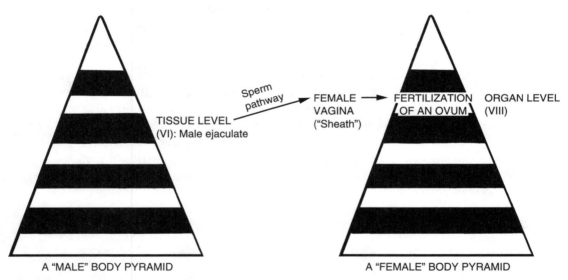

Fig. 13.3 Sperm ejaculation during intercourse: a connection between Male and Female Body Pyramids.

3. THE REPRODUCTIVE (FERTILIZED OVUM) PATHWAY AND ITS EXIT FROM THE PYRAMID

Fertile females experience a sudden surge in the blood concentration of luteinizing hormone (LH) about halfway through each of their *menstrual* (**MEN**-stroo-al) – "monthly" – *cycles*. As explained back in Chapter 8 (endocrine glands and their hormones), LH is one important type of trophic hormone secreted by the anterior pituitary gland.

Fertilization of an ovum

The luteinizing hormone circulates to one of the two *ovaries*. Here it dissolves and weakens the ovary's wall, such that ovulation – release of a mature ovum – occurs. Specifically, it is a *mature ovarian follicle* ("tiny sac" of secreting and nutrient-providing cells around the ovum) that ruptures under the influence of LH, thereby releasing the mature egg cell. A glance back at Figure 13.3 tells the basic story of fertilization. It involves the union or fusing together of a sperm nucleus with the ovum nucleus, when they meet in the outer third of an *oviduct* (**OH**-vih-**dukt**) or "egg tube."

The result of this fertilization process is a single cell called the *zygote* (**ZEYE**-goht). The zygote gets its name from the fact that it represents the "yoking together" of the nuclei of two cells, the sperm and the ovum.

The female reproductive pathway

Figure 13.4 shows what happens to the zygote after it is created. The zygote is essentially the very first stage of the embryo – the little body that "swells" and enlarges during the first 3 months after fertilization. As the zygote moves through the oviduct, it progressively divides by mitosis. This generates larger and larger masses of cells. First there is a *morula* (mor-**OO**-lah), a solid, "little mulberry"-like mass of cells. The morula keeps dividing and forms a *blastula* (**BLAS**-chew-lah), also called a *blastocyst* (**BLAS**-toh-**sist**).

The blastula (blastocyst) is a "little sprouter" or a "hollow sprouting bladder." These anatomic names reflect the fact that this early stage of the embryo is basically a sphere of cells which (like a bladder) contain a hollow cavity. The blastula is the stage that actually implants itself into the *endometrium* (**en**-doh-**MEE**-tree-um). The endometrium is the "inner" (*endo-*) layer of epithelial tissue that lines the "womb" (*metr*) or *uterus* (**YOO**-ter-**us**).

Further cell division of the implanted blastula leads to its *differentiation* – a "process of becoming different" or divided into separate masses of specialized body cells. This produces a *gastrula* (**GAS**-troo-lah) or hollow "little stomach" with three *primary germ layers* in its wall. These germ layers are collectively called *derms*, because they each represent a different "skin" or layer within the gastrula. These skins ultimately produce the four basic types of body tissues.

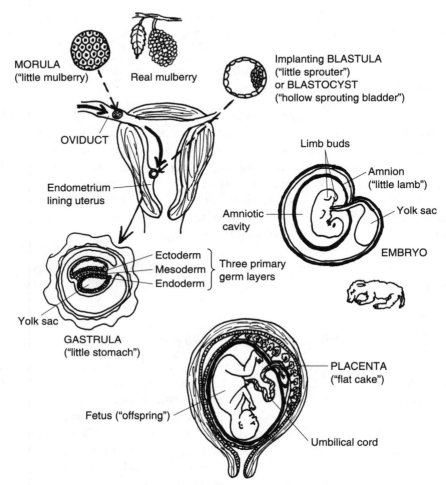

Fig. 13.4 Early stages in the female reproductive pathway.

Eventually, the rounded gastrula lengthens, then curls up into a recognizable, worm-like body with short *limb buds*. The growing embryo is now surrounded by the *amnion* (**AM**-nee-un), a thin membrane that protects it in the same way that it does unborn "little lambs" (*amni*).

From nine months after fertilization until the time of birth, the developing child is called a *fetus* (**FEE**-tus), or unborn "offspring." The fetus is nourished within the uterus by its *umbilical* (um-**BILL**-ih-kal) *cord*, which attaches to the "central pit" of its abdomen, the umbilicus. At the other end of the umbilicus is the *placenta* (plah-**SEN**-tah), named for its resemblance to a red "flat cake" (*placent*). The placenta is an organ of metabolic transfer of waste products and nutrients between the mother's bloodstream and that of her fetus.

The amnion around the fetus contains the *amniotic* (**am**-knee-**AH**-tik) *cavity*, which is filled with a clear, salty, *amniotic fluid*. For this reason, the amnion is sometimes called the "bag of waters," where the "waters" are, of course, the watery amniotic fluid.

At the end of the *gestation* (jes-**TAY**-shun) – child-"bearing" – period, the amnion ("bag of waters") finally ruptures. A *neonate* (**NEE**-oh-**nayt**) or "newborn" infant emerges from the vagina. In this case, the vagina serves an additional function as the so-called *birth canal*.

Summary of the female reproductive pathway

All of the preceding events can be collected together and described as the *female reproductive pathway*. This pathway begins with the zygote (fertilized ovum) and finally ends with the birth of the neonate through the vagina (birth canal).

The female reproductive pathway can be abstractly modeled (Figure 13.5) as a dark line moving out of the *organ* level of the Great *Female* Body Pyramid. This particular organ of neonate exit is, of course, the vagina or birth canal.

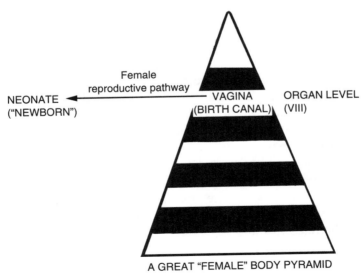

Fig. 13.5   Childbirth as a pathway moving out of the organ (vagina) level in the Great *Female* Body Pyramid.

Organism 1

THE FEMALE REPRODUCTIVE PATHWAY	=	PATHWAY OF THE ZYGOTE (FERTILIZED OVUM) AND ITS SUCCESSORS	=	A PATHWAY LEADING TO BIRTH (EXIT OF THE NEONATE) THROUGH THE VAGINA

Labor and Birth: A Beneficial Positive Feedback Cycle

Finally, after nine months of gestation, a neonate will be expelled from the birth canal of a pregnant woman by the *positive feedback process* of *labor and natural childbirth*.

Way back in Chapter 2, we introduced the concept of feedback control systems or cycles. As you may recall, a positive feedback system or cycle was considered an "out-of-control" system. [**Study suggestion:** If you don't remember very much about positive feedback systems, it would be a very good idea for you to review Chapter 2, before you go further on with this chapter.]

In Chapter 2, we did mention several examples where positive feedback cycles going to some dramatic climax were beneficial, rather than harmful, to the body. Sexual arousal leading to eventual climax or *orgasm* (**OR**-gazm), for instance, is a positive feedback cycle in the male that normally *must* occur if he is to ejaculate spermatozoa and fertilize an ovum. (Ejaculating sperm and getting a woman pregnant are pretty dramatic events!)

The other main example of beneficial positive feedback cited in Chapter 2 was contractions of the "womb" (uterus) that lead to childbirth. We will now examine this amazing process in more detail (see Figure 13.6).

Step 1: Estrogen increases irritability of the uterine smooth muscle to oxytocin. The best place to start is where the action is! This is the *myometrium* (**my**-oh-**MEE**-tree-um), a thick layer of smooth "muscle" (*myo-*) within the wall of the "uterus" (*metr*). The smooth muscle fibers of the myometrium encircle the uterus, so that when they contract, they constrict and powerfully push upon the fetus inside.

During the last few weeks of pregnancy, the very large *placenta* starts to dramatically increase its secretion of the hormone *estrogen* into the mother's bloodstream. Estrogen molecules circulate to the uterine wall. Here they stimulate the smooth muscle fibers in the myometrium to increase their

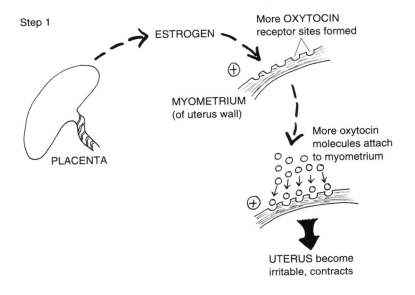

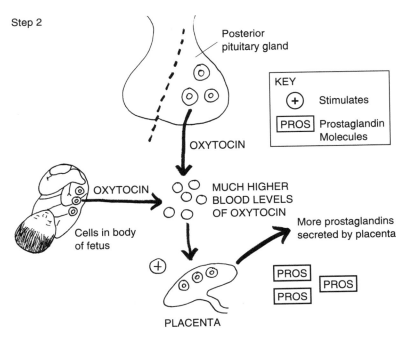

Fig. 13.6 Labor and childbirth; a positive feedback cycle. Step 1: Estrogen from the placenta increases the sensitivity of the uterus to oxytocin. Step 2: Greater release of oxytocin from posterior pituitary gland of the mother, and by the body of the fetus, stimulates more release of prostaglandins from the placenta.

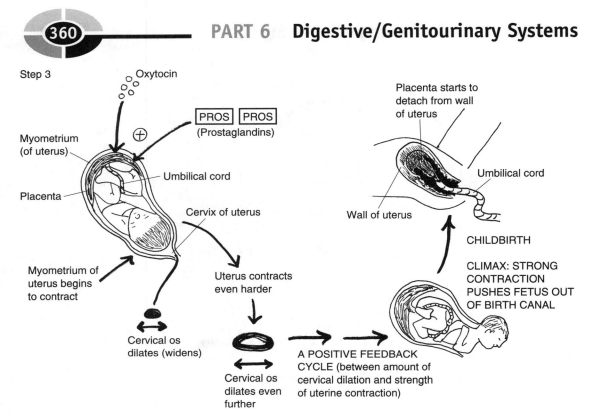

Fig. 13.6   (continued) Step 3: A positive feedback cycle between strength of uterine contraction and dilation of the cervix, eventually leads to childbirth.

formation of *oxytocin* (**ahk**-see-**TOH**-sin) *hormone receptor sites*. The name *oxytocin* literally means "a substance" (-*in*) that causes "swift" (*oxy-*) "birth" (*toc*). Oxytocin, then, is a hormone secreted by the hypothalamus, but stored and released by the posterior pituitary gland, which powerfully stimulates uterine smooth muscle to contract. With more oxytocin hormone receptor sites being produced due to the influence of estrogen, the myometrium binds (attaches) more oxytocin hormone molecules from the bloodstream. This makes the myometrium much more irritable, such that it may begin weak, on-and-off, uterine contractions.

Step 2: Oxytocin and prostaglandins begin a positive feedback uterine muscle contraction cycle. As birth gets very near, both the mother's posterior pituitary, as well as the fetus, itself, suddenly begin to release oxytocin into the bloodstream at a very rapid rate. The oxytocin in turn stimulates the placenta to increase its secretion of *prostaglandins* (**PRAHS**-tah-gland-ins). Once thought to be secreted only by the prostate gland (hence their name), the prostaglandins are a group of very active fatty acids that have a wide range of important biological effects. One of the most important effects of the prostaglandins is to act as a synergist with oxytocin, powerfully boosting its stimulating effect upon the contraction of the myometrium. Therefore, *oxytocin* and *prostaglandins* act together as the hormones that *start labor*.

Molecule 2

Step 3: A positive feedback cycle occurs between the amount of cervical dilation and the strength of uterine contraction. The myometrium (uterine smooth muscle) begins to forcefully contract due to strong stimulation by both oxytocin and prostaglandins. The uterus has a *cervix* (**SIR**-viks) – a tapered, inferior "neck" area. The *cervical os* is the tiny "mouth" at the tip of the uterus, which opens into the vagina or birth canal.

As the uterus begins to contract, the cervical os begins to dilate. And as the cervical os dilates, the uterus contracts even harder. A positive feedback cycle therefore kicks in, between the degree of dilation (widening) of the cervical os, and the strength of uterine smooth muscle contraction. The wider the cervical os dilates, the harder the uterus contracts. And the harder that the uterus contracts, the wider the os dilates. Finally, when the os is fully dilated, a strong uterine contraction propels the fetus out through the os, then out through the birth canal. The positive feedback cycle thus ends with a dramatic climax – the birth of a new human being!

Controlling the Output of Urine: The Urinary Excretion Equation

The *excretion* (E) of urine, like the process of labor and childbirth, is importantly influenced by several hormones. The job of urine formation is mainly done by the *nephrons* (nef-**RAHNS**), the microscopic structural and functional units of the kidney. (Consult Figure 13.7, A.)

Each nephron begins with a *glomerulus* (gluh-**MAHR**-yew-**lus**). The glomerulus is a tiny, red-colored collection of *renal capillaries*. This structure derives its name from its striking resemblance to a little red "ball of yarn" (*glomerul*). The blood pressure pushing against the inner walls of the capillaries in each glomerulus causes a *filtration* (F) of fluid out of the glomerulus, and into the adjoining group of *urinary tubules* – "tiny urine tubes."

Figure 13.7 (B) shows the processes involving the structures of each nephron in creating urine from the blood plasma. The summary equation expressing the formation of urine by the nephrons is:

$$E = F - R + S$$

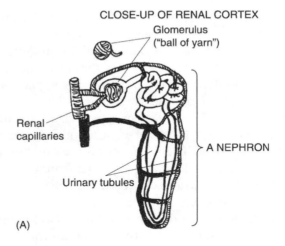

CLOSE-UP OF RENAL CORTEX

Glomerulus
("ball of yarn")

Renal
capillaries

A NEPHRON

Urinary tubules

(A)

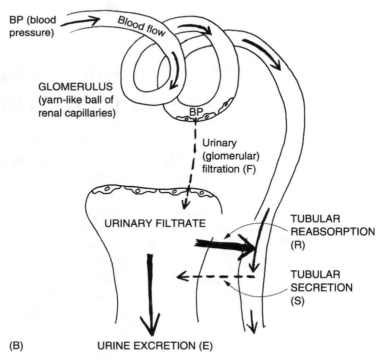

BP (blood
pressure)

Blood flow

GLOMERULUS
(yarn-like ball of
renal capillaries)

BP

Urinary
(glomerular)
filtration (F)

URINARY FILTRATE

TUBULAR
REABSORPTION
(R)

TUBULAR
SECRETION
(S)

(B)   URINE EXCRETION (E)

Fig. 13.7   The nephron and the critical processes involved in urine formation. (A) Anatomy
of the nephron: our urine-former. (B) The processes involved in excretion of urine.

In this equation, E stands for "excretion," F for the amount of blood "filtered," R for the amount of *tubular reabsorption*, and S for *tubular secretion*.

The urinary *filtrate* (**FIL**-trayt), or filtration product, comes from the pushing force of the blood pressure against the walls of the renal capillaries in the *glomeruli* (glah-**MEHR**-you-**lie**). This quantity of filtrate is huge, averaging about 180 liters of fluid per day in an average adult! [**Study suggestion:** If a person's blood volume is 6 liters, then how many times is the person's total blood volume filtered into the kidney tubules every single day?]

While it may seem wasteful, the huge volume of urinary filtrate (F) acts as the starting pool of fluid and solutes for what eventually becomes excreted as the urine. Because there is so much of this filtrate, the body can adjust many factors to influence how much urine is actually excreted, under particular physiological conditions.

After urinary filtration, one of the chief processes is tubular reabsorption (R). Reabsorption is the movement of material out of the filtrate, across the walls of the urinary tubules, and finally back into the bloodstream. Consider, for example, the tubular reabsorption of glucose. Under normal conditions, almost 100% of the glucose that is filtered into the kidney tubules is eventually reabsorbed back into the bloodstream. As a result, the urine excreted from the body is nearly free of glucose.

Under typical conditions, about 99%, or 179 liters, of the urinary filtrate (mostly water) is reabsorbed. [**Study suggestion:** If a person becomes extremely dehydrated, as after excessive sweating, then what do you predict will happen to the amount of H_2O reabsorbed? Will the percent (%) reabsorbed increase above typical amounts, or decrease below it? Why?]

Another process, tubular secretion (S), involves the active (ATP-requiring) addition of small quantities of particular chemicals from the bloodstream into the urinary tubules. Molecules of *penicillin* (pen-ih-**SILL**-in) and many other *antibiotics* (**an**-tih-buy-**AH**-ticks), for instance, are just too large to be filtered across the walls of the glomeruli. Hence, the epithelial cells lining the blood vessels actively pump the penicillin out of the bloodstream and into the urinary tubules. Therefore, penicillin is excreted (E) out of the body, via the urine. Other substances that can be filtered across the glomerular wall, but are still present in huge quantities within the blood plasma, can also be secreted into the urine. An example of this are the millions of H^+ (acid) ions removed from the bloodstream by tubular secretion. Although only a few milliliters of total fluid are generally secreted each day, they still have a significant influence.

Summarizing our previous urinary excretion equation and plugging in some numbers, we obtain:

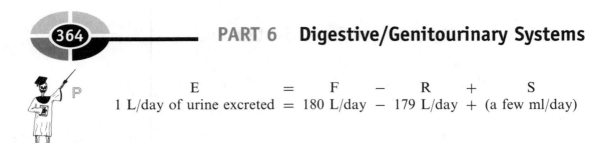

$$E \qquad = \quad F \quad - \quad R \quad + \quad S$$
$$\text{1 L/day of urine excreted} = \text{180 L/day} - \text{179 L/day} + \text{(a few ml/day)}$$

Organ System 1

Body Water Balance and ADH

Since urine primarily consists of water, the percentage of water that is reabsorbed back into the bloodstream can have giant effects upon the volume of urine that is eventually excreted. The *body water balance* between the amount of water reabsorbed and kept in the bloodstream, and the amount of water flowing out into the urine, is critically dependent upon a single hormone. This hormone is antidiuretic hormone (ADH).

We have met ADH earlier, within Chapter 8 on the endocrine glands. It is secreted by the hypothalamus, then stored and released intermittently by the posterior pituitary gland. By the word antidiuretic, we mean something that is literally "against" (*anti-*) "diuresis" (**die**-ur-**EE**-sis). In plain English, ADH is against excessive urination.

ADH works primarily by increasing the amount of water (H_2O) being reabsorbed out of the kidney tubules, and back into the bloodstream by osmosis. To help us understand the general effect, consider a couple of simple models (Figure 13.8). We will represent the excretion (E) of urine from the kidney as a gigantic faucet.

Condition 1: Low blood [ADH]. When ADH is absent or present at very low concentration, the excretion of urine from the kidney "faucet" is gushing water at full blast. The faucet is considered fully open, because there is nothing reabsorbing the H_2O (holding the water back). The vast 180 liters of urinary filtrate produced every day are mostly excreted. There is an uncontrolled diuresis (excessive amount of urine excreted). Before long, a severe tissue dehydration – drying out due to lack of water – occurs. Blood volume and blood pressure dramatically drop, and the person may collapse into a coma and die!

A real clinical example of this problem is provided by the disease called *diabetes* (**die**-uh-**BEE**-teez) *insipidus* (in-**SIP**-ih-**dus**) – a watery (and therefore "tasteless") excessive "passing through" of urine. The person suffering from diabetes insipidus has so much water left in his urine that he may excrete *10* liters per day, instead of the normal *1*! Severe dehydration due to extreme watery diuresis soon follows. The causes may include trauma to the head, which damages the sources of ADH (the hypothalamus and posterior pituitary gland).

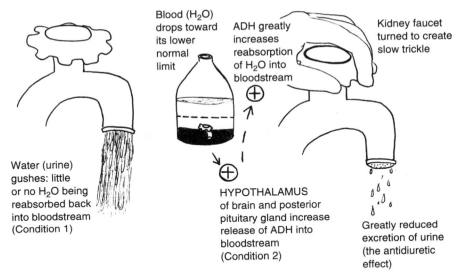

Blood (H₂O) drops toward its lower normal limit

ADH greatly increases reabsorption of H₂O into bloodstream

Kidney faucet turned to create slow trickle

Water (urine) gushes: little or no H₂O being reabsorbed back into bloodstream (Condition 1)

HYPOTHALAMUS of brain and posterior pituitary gland increase release of ADH into bloodstream (Condition 2)

Greatly reduced excretion of urine (the antidiuretic effect)

Fig. 13.8   Two giant body water faucets: with and without ADH. (1) Low blood [ADH]: diabetes insipidus and an uncontrolled high volume of urine excretion from the wide-open kidney "faucet." (2) Normal blood [ADH]: as blood [H₂O] falls toward its lower normal limit, the hypothalamus and posterior pituitary gland increase their release of ADH; greatly increased reabsorption of water turns kidney faucet down to excreting urine at a slow drip.

Condition 2: Normal blood [ADH]. Suppose that ADH is being secreted in normal amounts. When a giant kidney faucet is operating at full blast, the level of water in the big jug (body fluid content) attached to the faucet soon begins to run low. (This models a drop in the blood H₂O concentration toward its lower normal limit.) Such a situation provides a stimulus for greater ADH secretion.

The ADH hormone molecules can be visualized as a big hand that strongly turns the giant kidney faucet down to a slow trickle. This dramatic reduction in urine excretion will tend to raise the blood H₂O concentration back toward its set-point level. The fluid reservoir attached to the faucet does not run down too low. Therefore, tissue dehydration is prevented.

ADH SUMMARY

In summary,

Molecule 3

	A	D	H
means	Anti-	*diuretic*	hormone
	"Against"	"Excessive *urination*"	

and, therefore, might also be called:

	Anti-	*dehydration*	hormone
	"Against"	"Excessive body	
		tissue *drying out*"	

Body Salt (Electrolyte) Balance

In addition to body water balance, body *salt or electrolyte balance* is likewise critical for human survival. To be specific, the concentration of sodium (Na^+) ions within the bloodstream is vital. Why? Remember that sodium ions (in addition to calcium, Ca^{++} ions) are the main chemical particles responsible for neuromuscular excitation! Upsets in their balance (too high or too low an ion or electrolyte concentration) therefore affect how nerve fibers excite muscles to contract, how well we can think, and even whether our heart continues to beat or not!

NORMONATREMIA THE IDEAL

The ideal state of affairs for maintaining normal, healthy levels of neuromuscular excitability, then, is a maintenance of *normonatremia* (**nor**-moh-nah-**TREE**-me-uh). Translated into everyday English, normonatremia is a "normal" (*normo-*) "blood concentration of" (*-emia*) "sodium" (*natr*). With conditions of normonatremia, our heart muscle, neurons, and other ionically excitable tissues usually operate smoothly and with great cooperation and coordination.

THE DEADLY THREAT OF HYPONATREMIA

"What happens if I work out and lose lots and lots of sodium in my *sweat*, Professor?"

In that case, Baby Heinie, we would be looking for *renal compensation*, or "kidney payback," for help in correcting such a large Na^+ loss. Remember that urine is a very salty fluid, with a high NaCl content. So we need a mechanism to increase the tubular reabsorption (R) of sodium ions from the urinary filtrate.

"Well, what if I just keep sweating and sweating, don't take any salt pills, and don't reabsorb any more Na^+ ions?"

In this case, you may well end up with *hyponatremia* (**high**-poh-nah-**TREE**-me-ah) – "a blood condition of" (*-emia*) "deficient or below-normal" (*hypo-*) "sodium" (*natr*) concentration. Now, why should you be concerned about suffering from hyponatremia, Baby Heinie? [**Study suggestion:** From your current reading, why do *you* think that Baby Heinie should be worried about suffering from hyponatremia?]

Tissue 1

ALDOSTERONE: THE SALT-CONSERVING HORMONE

Renal compensation for excessive body Na^+ loss usually comes in the form of an increased secretion of *aldosterone* (al-**DAHS**-ter-ohn). The hormone, aldosterone, is a steroid secreted by the adrenal cortex. The stimulus for its secretion is a drop in the blood sodium ion concentration toward its lower normal limit (as after a bout of excessive sweating).

When the appropriate gland cells in the adrenal cortex are stimulated (through an indirect process), they increase their secretion of aldosterone into the bloodstream. Aldosterone functions by increasing the reabsorption of Na^+ ions from the urinary filtrate in the kidney tubules, back into the bloodstream. [**Study suggestion:** Think of a giant hand with the word, **aldosterone**, written on the back of it. The hand reaches down into the urinary filtrate (wearing a glove, of course) and grabs bunches of salt crystals, pulling them out of the filtrate before it goes down the drain as urine.]

Molecule 4

As a result, the blood $[Na^+]$ usually rises back up toward its set-point level, and the threat of hyponatremia is prevented. In summary, we can think of aldosterone as our main "salt-conserving" hormone.

A FINAL GOOD-BYE

Congratulations! You have now finished reading *PHYSIOLOGY DEMYSTIFIED*! (Have you checked out *ANATOMY DEMYSTIFIED*, its *pairmate*?) "Now, *goodnight*, Baby Heinie!" "*Goodnight* and *thanks*, Professor Joe!"

Quiz

Refer to the text in this chapter if necessary. A good score is at least 8 correct answers out of these 10 questions. The answers are listed in the back of the book.

1. The G part of the G-U system is primarily involved in the function of:
   (a) Urinary excretion
   (b) Respiration
   (c) Reproduction
   (d) Blood pressure

2. The main nitrogen-containing solute found within the urine:
   (a) Urea
   (b) NaCl
   (c) Ammonia
   (d) Peptides

3. The urinary pathway leaves the Great Body Pyramid at the _____ level:
   (a) Organelle
   (b) Cell
   (c) Organ
   (d) Chemical

4. The end-body or cap on the head of a sperm:
   (a) Acrosome
   (b) Chromosome
   (c) Mitochondrion
   (d) Flagellum

5. The fluid added to the sperm when they are ejaculated:
   (a) Blood serum
   (b) Lymph
   (c) Semen
   (d) Bilirubin

6. The actual body structure that releases an ovum for possible fertilization:
   (a) Penis
   (b) Fallopian tube
   (c) Urethra
   (d) Mature follicle

7. The child-bearing period:
   (a) Ejaculation
   (b) Gestation
   (c) Emasculation
   (d) A long vacation!

8. Two substances critical for starting labor:
    (a) ADH and aldosterone
    (b) Insulin and testosterone
    (c) Estrogen and progesterone
    (d) Prostaglandins and estrogen

9. The muscular wall of the uterus:
    (a) Myometrium
    (b) Dorsal venous sinus
    (c) Endometrium
    (d) Corpus striatum

10. The salt-conserving hormone:
    (a) Vasopressin
    (b) ADH
    (c) Insulin
    (d) Aldosterone

Body-Level Grids for Chapter 13

Anatomy and *Biological Order* **Fact Grids for Chapter 13:**

TISSUE
Level

1

ORGAN SYSTEM
Level

1

Anatomy and *Biological Disorder* Fact Grids for Chapter 13:

TISSUE
Level

1

Physiology and *Biological Order* Fact Grids for Chapter 13:

MOLECULE
Level

1	2

3	4

ORGAN SYSTEM
Level

1

ORGANISM
Level

1

Physiology and *Biological Disorder* Fact Grids for Chapter 13:

P

ORGAN
Level

1

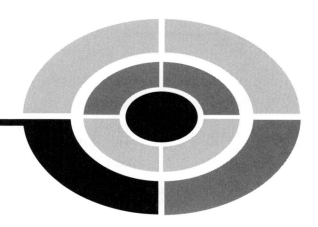

Test: Part 6

DO NOT REFER TO THE TEXT WHEN TAKING THIS TEST. A good score is at least 18 (out of 25 questions) correct. Answers are in the back of the book. It's best to have a friend check your score the first time, so you won't memorize the answers if you want to take the test again.

1. In Ancient Rome, fasces were:
   (a) Dregs
   (b) Defecated material
   (c) Salivary glands
   (d) Cheek cell smears
   (e) Bundles of brown sticks

2. The release of some useful substance into the lumen of the digestive tract:
   (a) Reabsorption
   (b) Secretion
   (c) Digestion
   (d) Ingestion
   (e) Desiccation

3. Hepatocytes would be found in the:
   (a) Liver
   (b) Skeletal muscles
   (c) Duodenal wall
   (d) Dirty wash
   (e) Spleen

4. Carries bile into the gall bladder:
   (a) Cystic duct
   (b) Jejunal fold
   (c) Serosa lumen
   (d) Common bile duct
   (e) Pancreatic duct

5. Sphincter guarding entrances of both the common bile duct and pancreatic duct:
   (a) Cardiac
   (b) Lower esophageal
   (c) Internal anal
   (d) Hepatopancreatic
   (e) Biliary

6. CCK represents a:
   (a) Symbol for a liquid detergent substance
   (b) Type of enzyme found in RBCs
   (c) Hormone that stimulates the release of stored bile
   (d) Chemical messenger that promotes gastric secretions
   (e) Species of intestinal parasites

7. Results from a state of *overhydration*:
   (a) Low blood [ADH]
   (b) Constipation
   (c) Joint crystals
   (d) High blood [ADH]
   (e) High blood [aldosterone]

8. Secretion is inhibited whenever duodenal pH rises beyond 7:
   (a) Insulin
   (b) Secretin
   (c) Glycogen
   (d) Amylase
   (e) Starch

9. Proteases have as their main targets:
    (a) Protein fragments
    (b) Carbohydrate polymers
    (c) Simple sugars
    (d) Membrane lipids
    (e) Cell mitochondria

10. The final products of fat digestion, which are absorbed:
    (a) Amino acids such as glycine
    (b) Chunks of chyme
    (c) Frozen slush
    (d) Fatty acids and glycerol
    (e) Carbonic anhydrase

11. Greatly increase surface area for absorption within the small intestine:
    (a) Rugae
    (b) Submucosae
    (c) Villi and microvilli
    (d) Gastric enzymes
    (e) Mucosal peptides

12. Thought to be mainly responsible for excitation of the GI tube wall:
    (a) Longitudinal smooth muscle
    (b) Goblet cells
    (c) Circular smooth muscle
    (d) Tunica adventitia
    (e) Cholecyst

13. Constriction around a single point of the digestive tube:
    (a) Mass movements
    (b) Duodenal stretching with chyme
    (c) Vasoconstriction
    (d) Valsalva's maneuver
    (e) Peristalsis

14. Produces thorough mixing of chyme with digestive enzymes:
    (a) Segmentation
    (b) Gastric bypass
    (c) Cholecystectomy
    (d) Sphincter closure
    (e) Appendectomy

15. Mainly responsible for coloring urine an amber hue:
    (a) Urobilin

(b) Acetyl CoA
(c) Lysosomes
(d) Gut bacteria
(e) Ammonia

16. According to the Preformation Theory:
    (a) Sperm cells contained homunculi
    (b) Mature ova sometimes talked amongst themselves
    (c) Eggs and sperm cells were already formed in the blastula
    (d) People were born with tails, then lost them
    (e) Spermatozoa were always ejaculated as pre-formed pairs

17. Ejaculated spermatozoa have a life expectancy of about _____ hours:
    (a) 2
    (b) 48–72
    (c) 100–300
    (d) 5,000–10,000
    (e) 10 years

18. Makes up about 60% of the semen:
    (a) Salivary exudate
    (b) Seminal vesicle fluid
    (c) Prostate fluid
    (d) Female vaginal mucus
    (e) Spermatozoa

19. Represents the "yoking together" of a sperm plus ovum:
    (a) Morula
    (b) Blastocyst
    (c) Ectoderm
    (d) Zygote
    (e) Chromosome bead

20. The "bag of waters":
    (a) Amnion
    (b) Umbilical cord
    (c) Blastula
    (d) Urocyst
    (e) Dura mater

21. A tubal ligation essentially interrupts:
    (a) Photosynthesis
    (b) Spermatogenesis
    (c) Cell division

(d) The female reproductive pathway

(e) Entry of a sperm into the vagina

22. Labor and childbirth are good examples of:
    (a) Negative feedback systems
    (b) Homeostatic mechanisms
    (c) Cybernetic cycles
    (d) Linear programs
    (e) Positive feedback

23. Starts secreting large amounts of estrogen during the final weeks of pregnancy:
    (a) Dermis
    (b) Umbilical cord
    (c) Liver
    (d) Placenta
    (e) Stomach

24. Smooth muscle portion of the uterine wall:
    (a) Endometrium
    (b) Glandular epithelium
    (c) Renal capsule
    (d) Mesoderm
    (e) Myometrium

25. Tiny yarn-like ball of renal capillaries:
    (a) Collecting duct
    (b) Glomerulus
    (c) Transverse colon
    (d) Cardia
    (e) Genital tubercle

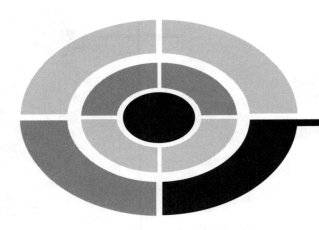

Final Exam

DO NOT REFER TO THE TEXT WHEN TAKING THIS EXAM. A good score is at least 75 correct. Answers are in the back of the book. It's best to have a friend check your score the first time, so you won't memorize the answers if you want to take the test again.

1. Speaking broadly, an organism is:
   (a) A collection of unrelated organs and organ systems
   (b) Some multi-celled creature with no apparent design
   (c) Any living body with a high degree of recognizable pattern
   (d) Several associated body structures whose functions are unrelated
   (e) One or more apparently randomly constructed bodies

2. To many early Greek thinkers, the human body was a microcosm. This implies that:
   (a) Living things do not appear to follow fixed laws
   (b) The internal environment is a tiny representative of the external environment
   (c) A macrocosm existed that directly contradicted whatever happened at the microcosm

    (d) The Laws of Nature were apparently suspended when humans were around

    (e) No larger Universe exists

3. First coined the word physiology:
    (a) Hippocrates
    (b) Galen
    (c) Aristotle
    (d) Claude Bernard
    (e) Socrates

4. Plain functions are:
    (a) Actions that non-living structures in the body perform
    (b) Structures that occupy space
    (c) Best represented by nouns rather than verbs
    (d) Whatever all living or non-living bodies do
    (e) Seldom apparent to outside observers

5. A roughly S-shaped pattern of relative constancy:
    (a) Anatomy
    (b) Hemostasis
    (c) State of chaos
    (d) Morbidity
    (e) Homeostasis

6. According to the Law of Complementarity:
    (a) We should give each other compliments
    (b) Metabolism is a "state of fixity"
    (c) The existence of body structures having particular characteristics limits or determines the possible functions those structures might have
    (d) Oral body temperature is absolutely constant
    (e) "Monkey see, monkey do"

7. A stimulus is:
    (a) Something that inhibits the activities of a sensory receptor
    (b) A group of nerve impulses that always result in a corrective action
    (c) Any creature that causes others to withdraw
    (d) A detectable change in the body's internal environment, or within the external one
    (e) Generally preceded by a response

8. Pathophysiology represents a:
    (a) Highly orderly progression of changes in body structure

(b) Change in body structure that reflects normal body function

(c) State wherein physiology is diseased and greatly abnormal

(d) Type of untreatable medical condition

(e) Bit of order within an otherwise diseased organism

9. A Great Pyramid of Structure-Function *Dis*order would involve:

(a) Regular arrangements of body structures to perform functions that were also regular

(b) Loss of pattern between body structures and their functions

(c) Maintenance of homeostasis across various organ systems

(d) Stacked grids of body structures that fit nicely into healthy groups

(e) Many levels of biological organization that interacted normally with one another

10. Physiology is limited to these levels of body organization:

(a) Atoms and subatomic particles

(b) Organ systems and everything below them

(c) Living cells and everything above them

(d) Molecules, but not atoms

(e) Most subatomic particles (except for electrons)

11. The concept of a lower normal limit (LNL) is valuable because:

(a) Accurate verbal descriptions of people and processes are obtained

(b) Sensitivity to the personal feelings of the subject are observed

(c) It provides a floor, below which the parameter is in a state of deficiency

(d) Qualitative thinking subject to verification and checking by others is attained

(e) It provides a ceiling, above which the parameter is in a state of excess

12. Parameters:

(a) Are nothing but recorded data

(b) Seldom provide reliable information

(c) Are produced by measuring and recording the results in certain numbers of units

(d) Are the same things as body structures

(e) Frequently involve functions that cannot be tied down

13. Shoe weight in pounds:

(a) Anatomical parameter

(b) Physiology

(c) Function

(d) Structure

(e) Structural parameter

14. An aspect of living body function that has been measured and expressed in a certain number of units:
    (a) Structural parameter
    (b) Feedback loop
    (c) Functional parameter
    (d) Negative change
    (e) Physiological parameter

15. The book said that the phrase "Compu-think" can be interpreted to mean:
    (a) Computer-like modes or ways of human thinking
    (b) An over-reliance upon technology is foolish
    (c) Science is a bad thing
    (d) More attention should be given to arts in the school curriculum
    (e) Teachers should be paid higher salaries

16. "Regression towards the mean," or "things usually tend to balance-out":
    (a) A hypo-state
    (b) Set-point value
    (c) Diabetes insipidus
    (d) Upper normal limit
    (e) Random "noise" within an otherwise controlled environment

17. Milliliters (ml) and liters (L) are typical units of:
    (a) Mass or weight
    (b) Concentration and density
    (c) Volume
    (d) Area
    (e) Length

18. The typical units for recording blood pressure (BP):
    (a) Degrees Fahrenheit
    (b) Cubic centimetres
    (c) Grams per liter of blood
    (d) Millimeters of mercury (Hg)
    (e) pH units

19. The ancient Humoral Doctrine was somewhat like the modern concept of:
    (a) Absolute constancy

(b) Balance as a rough equality between two or more different things

(c) Strict following of laws that are essentially meaningless

(d) An imbalance between the Universe and ourselves

(e) Wisdom is found in the details!

20. The Father of Modern Medicine:
(a) Basel
(b) Plato
(c) Vesalius
(d) Hippocrates
(e) Galileo

21. An upside-down, dead giraffe without spots is most closely modeling:
(a) Pathophysiology
(b) Normal function
(c) Homeostasis
(d) Negative feedback
(e) Biological order

22. Provided the first strong evidence of natural chemical synthesis in the body, by performing experiments on living animals:
(a) Barnum & Bailey
(b) Schleiden & Schwann
(c) Claude Bernard
(d) William Harvey
(e) Sir Robin of Loxley

23. The phrase *milieu intérieur* comes from the French for:
(a) "Nourishing substances"
(b) "A state of change"
(c) "External surroundings"
(d) "The raging sea"
(e) "Internal environment"

24. The fluid lying outside of our cells:
(a) Organic paste
(b) Intracellular fluid
(c) Organelles
(d) Extracellular fluid
(e) Solid rock

25. Originated the concept of a "fight-or-flight" response:
(a) Walter Cannon
(b) Claude Bernard

(c) Ernest Starling
(d) Hippocrates
(e) Connie Chung

26. First used the term homeostasis:
    (a) Walter Mondale
    (b) Galen
    (c) Leonardo da Vinci
    (d) Albrecht Dürer
    (e) Walter Cannon

27. The use of parameters generally improves the quality of information regarding homeostasis, because:
    (a) More vagueness of definition allows more ease in translation
    (b) Structure determines function (except when both are being measured)
    (c) Whether a relative constancy is being maintained can be directly assessed
    (d) They provide less information, but it is more honest
    (e) Some people can't be trusted!

28. Cybernetics is the:
    (a) Ancient art of translating writing
    (b) Modern science of communication and control in man and machine
    (c) Relative constancy of thought waves
    (d) Same thing as kinetics
    (e) Opposite of control system theory

29. For a pendulum swinging from a clock, the set-point is:
    (a) A way of reading the time
    (b) The far-left swing
    (c) The far-right swing
    (d) The average position of the pendulum over time
    (e) The place where the pendulum swings so far that it breaks

30. Stressors:
    (a) Are exactly the same things as stimuli
    (b) Can effectively be removed from the environment
    (c) Occupy space, have mass and weight, and assume a definite form
    (d) Tend to disturb parameters from their set-point levels
    (e) Can always be corrected or removed

31. System whose output or response curves back upon the start to inhibit further activity of the system:
    (a) Negative feedback
    (b) Linear
    (c) Summative
    (d) Partially movable
    (e) Positive feedback

32. An afferent pathway:
    (a) Carries information away from a control center
    (b) Means a motor channel
    (c) Is generally a sensory pathway
    (d) Acts like a boomerang
    (e) Is just an efferent pathway

33. Establishes 98.6 degrees F as the set-point for oral body temperature:
    (a) Effector
    (b) Free nerve endings in the skin
    (c) Response
    (d) Stimulus
    (e) Control center

34. An "out-of-control" system:
    (a) Anatomical
    (b) Homeostatic
    (c) Normal immune
    (d) Positive feedback
    (e) Neutral

35. Body temperature way below the lower normal limit:
    (a) Normothermia
    (b) Hypotension
    (c) Euthyroidism
    (d) Hyperthermia
    (e) Hypothermia

36. A system that tends to maximize its own errors:
    (a) Blood glucose control
    (b) Vicious cycle
    (c) Benign line
    (d) Cybernetic
    (e) Missile guidance

37. If a state of homeostasis exists, then clinical health:

(a) Is absent
(b) Probably also exists
(c) Is totally unaffected
(d) May never come into being
(e) Has been irreversibly lost

38. A beneficial example of positive feedback:
(a) Climbing a tree higher and higher
(b) Singing so loudly that no one can stand it!
(c) Washing the dishes less, the more that someone complains
(d) Firing and transmission of a neuron action potential
(e) Blood pressure getting so low that a person faints

39. The Chemical Level of body organization:
(a) Atoms, but not cells
(b) Atoms, molecules, electrons, protons, and neutrons
(c) Mitochondria and cell nuclei
(d) Heart and cardiac muscle tissue
(e) Soup to nuts

40. The class of molecules rich in carbon (C) atoms:
(a) Crystals
(b) Electrolytes
(c) Inorganic
(d) Hydrated
(e) Organic

41. The chemical bond between H and O in H_2O:
(a) Polar covalent
(b) Ionic
(c) Non-electric
(d) Nonpolar covalent
(e) Double

42. Cations:
(a) Negatively charged ions
(b) Nonpolar solutes
(c) Positively charged ions
(d) Peptide bonds
(e) The same thing as anions

43. Anabolism is:
(a) Synthesis or build-up
(b) Exchange

    (c) Enzyme reaction

    (d) Breakdown or decomposition

    (e) Replacement

44. $HCl + NaHCO_3$:

    (a) $H^+ + Na_2CO_3$

    (b) $NaCl + H_2CO_3$

    (c) $H_2SO_4 - HCl$

    (d) $CO_2 + H_2O$

    (e) ATP + Phosphate

45. H^+ ion acceptors:

    (a) Bases (alkali)

    (b) H_2CO_3

    (c) HCl

    (d) Acids

    (e) Neutral substances

46. Has a low pH, but a high $[H^+]$:

    (a) Baking soda

    (b) Albumen

    (c) Vinegar

    (d) Blood

    (e) Saliva

47. Average or set-point blood pH level:

    (a) 4.8

    (b) 6.2

    (c) 7.3

    (d) 7.4

    (e) 7.5

48. Baby Heinie drinks a bottle of lye. He would likely suffer:

    (a) Hallucinations

    (b) Metabolic alkalosis

    (c) Delusions of grandeur

    (d) Metabolic acidosis

    (e) Neutral liquid overdose

49. Microphysiology can also be legitimately called:

    (a) Organelle physiology

    (b) Cell anatomy

    (c) Microbial fine structure

    (d) Cell functions

(e) Molecular interactions

50. Fluid mosaic:
    (a) Describes nature of the blood plasma
    (b) Structure of plasma membrane
    (c) The intercellular protein matrix
    (d) Pretty colored pictures on the body wall
    (e) Cell membrane protein bilayer

51. "Like dissolves like" means:
    (a) Oil and water don't mix!
    (b) Soap, like car wash liquid, can clean family pets
    (c) NaCl dissolves in $CH_3CH_2CH_3$
    (d) Nonpolar covalent substances bond with polar ones
    (e) Ionic compounds dissolve in hydrocarbon solvents

52. Glucose enters most cells by this process:
    (a) Osmosis
    (b) Facilitated diffusion
    (c) Simple diffusion through pores
    (d) Active transport
    (e) Phagocytosis

53. The ATP–ADP cycle:
    (a) Involves the splitting of ADP by ATPase
    (b) The combination of phosphate with ADP
    (c) Glucose combines with phosphorus
    (d) ATP sops up free energy and becomes AQP
    (e) H_2O and H_2CO_3 = ATP

54. Cellular respiration occurs in this part:
    (a) Cytoplasm
    (b) Lysosome
    (c) Glycogen droplet
    (d) Mitochondrion
    (e) ER

55. A single gene provides the:
    (a) Chemical code for the synthesis of a certain protein
    (b) List of ingredients for making an entire new organism
    (c) Structure of a whole DNA double helix
    (d) Codons for a particular lipid
    (e) Basis for absorbing most alcohols into the liver cells

56. The main dense fibrous connective tissue portion of the skin:
    (a) Hypodermis
    (b) Dermis
    (c) Hair follicles
    (d) Sebaceous glands
    (e) Subdermis

57. Blood plasma and other parts of the ECF are normally _____ solutions:
    (a) Hypertonic
    (b) Salt and hemoglobin
    (c) Isotonic
    (d) Keratin
    (e) Hypotonic

58. Cells that create the bone matrix:
    (a) Endothelial
    (b) Simple squamous epithelial
    (c) Neurons
    (d) Muscle fibers
    (e) Osteoblasts

59. Hypocalcemia is very dangerous, because:
    (a) Too little calcium in the urine may cause kidney stones
    (b) Potassium is also found at high concentrations within most cells
    (c) Excessive calcium may dull the sense of taste, leading to starvation
    (d) Normocalcemic states may just be taken for granted
    (e) Adequate blood Ca^{++} levels are absolutely necessary for muscles to contract

60. The major body function performed by red bone marrow:
    (a) Transmission of action potentials
    (b) Hematopoiesis
    (c) Thermoregulation
    (d) Water reabsorption
    (e) Na^+ excretion

61. In a bone–muscle lever system, the _____ serves as a lever:
    (a) Joint
    (b) Bone
    (c) Skin
    (d) Skeletal muscle

    (e) Blood vessel

62. The major central portion of a neuron, which contains its nucleus and most other organelles:
    (a) Soma
    (b) Dendrite
    (c) Axon
    (d) Synapse
    (e) Vesicles

63. An alternative name for the neuromuscular junction is the:
    (a) Axon hillock
    (b) Transmitter binding sites
    (c) Motor end-plate
    (d) Myofibril
    (e) Synapse

64. Within muscle fibers, the lateral sacs of the sarcoplasmic reticulum serve as:
    (a) Sliding filaments between molecules
    (b) Passageways for excitation
    (c) Terminals that release neurotransmitter molecules
    (d) H^+ ion acceptors
    (e) Ca^{++} ion storage depots

65. Myosin cross-bridges:
    (a) Places where actin changes into myosin
    (b) Chief points of contact between thin actin and thick myosin myofilaments
    (c) Are another form of myosin ATPase enzyme
    (d) Water-filled channels through muscle fibers
    (e) The same thing as sarcomeres

66. Troponin–tropomyosin serves to:
    (a) Attach a muscle fiber to a motor neuron
    (b) Stimulate fibers across the neuromuscular junction
    (c) Open up gaps within neuron cell membranes
    (d) Inhibit muscle contraction by preventing a chemical connection between actin and myosin
    (e) Excite or agitate ribosomes to synthesize more proteins

67. Electrical shock not producing any observable response from a muscle:
    (a) Maximal stressor

(b) Subthreshold stimulus
(c) Threshold prod
(d) Twitch
(e) Sarcomere

68. Either all of the muscle fibers in a given motor unit contract together at the same time, or none of them contract:
(a) Summation
(b) Recruitment of motor units
(c) The All-or-None Law
(d) Threshold
(e) Tetanus

69. Most continuous body movements represent these types of muscle contractions:
(a) Tetanus
(b) Fatigue
(c) Single twitches
(d) Inverted twitches
(e) Summations of single twitches

70. Serves as a reserve muscle energy storage, that can quickly be converted to ATP:
(a) Creatine phosphate
(b) Myoglobin
(c) Heme iron
(d) Calcium ions
(e) Myosin ATPase

71. Low oxidative (LO) muscle fibers are basically:
(a) Fast twitch
(b) Aerobic
(c) Slow twitch
(d) Myosin-deficient
(e) Anaerobic

72. A narrow, fluid-filled gap, where two neurons almost (but not quite) have their axons and cell body touch:
(a) Tight junction
(b) Axon hillock
(c) Synapse
(d) Nephron
(e) Ganglion

73. Myelinated nerve fibers:
    (a) Usually have a dull, grayish appearance
    (b) Saltatory conduction
    (c) Do not conduct action potentials at all
    (d) Local circuit currents
    (e) Intercalated discs

74. The automatic pulling back of some body part from a harmful or annoying stimulus:
    (a) Ventral rooting of a spinal nerve
    (b) Effector activation
    (c) Withdrawal reflex
    (d) Sensory pathway
    (e) Descending nerve tracting

75. "Ductless" glands of internal secretion:
    (a) Mixed
    (b) Salivary
    (c) Exocrine
    (d) Mammary
    (e) Endocrine

76. A given hormone only affects certain cells within the body, those having hormone receptor sites of appropriate shape and size:
    (a) Concept of a "Second Messenger"
    (b) "A little dab will do you"
    (c) Endocrine glands are leaky faucets
    (d) Hormones are "arousers"
    (e) Target cell specificity

77. Releasing hormones are secreted from the:
    (a) Anterior pituitary
    (b) Thyroid gland
    (c) Adrenal cortex
    (d) Hypophysis
    (e) Hypothalamus

78. The Great Mother Artery for the systemic circulation:
    (a) CPA
    (b) RA
    (c) Aortic arch
    (d) R pulmonary artery
    (e) Brachial artery

79. One complete sequence of systole and diastole of all four chambers of the heart:
    (a) "Lubb"
    (b) Cardiac Cycle
    (c) "Dup"
    (d) Auscultation response
    (e) Valve regurgitation

80. The normal EKG wave pattern:
    (a) M, S, Q, T
    (b) RS, VS, TPR
    (c) P, QRS, T
    (d) EDV, SV, RA
    (e) T, MAD, IVP

81. The amount of blood ejected from the ventricles with each beat of the heart:
    (a) CO
    (b) HR
    (c) SBP
    (d) SV
    (e) DBP

82. A residual or "left-over" blood pressure due to the force of snapping back of the stretched brachial artery wall:
    (a) Arteriosclerosis
    (b) Cardiac output
    (c) Blood pressure gradient
    (d) Diastolic BP
    (e) Systolic BP

83. A clear, watery filtrate of the blood plasma:
    (a) Serum
    (b) Bile
    (c) Pancreatic juice
    (d) Bicarbonate
    (e) Lymph

84. The most common formed element in the bloodstream:
    (a) Erythrocyte
    (b) Neutrophil
    (c) Platelet
    (d) Lymphocyte

(e) Globulin

85. The plasma cells have the important immune function of:
    (a) Presenting surface antigens
    (b) Releasing toxic chemicals that kill invading bacteria
    (c) Production of antibodies
    (d) Secretion of thymosin
    (e) Promotion of blood clotting

86. A lady with "A-positive" blood type would have _____ on her RBCs:
    (a) B agglutinogen, no Rh factor present
    (b) AB agglutinogens, Rh factor present
    (c) O agglutinogen, no Rh factor present
    (d) A agglutinogen, Rh factor present
    (e) A agglutinogen, no Rh factor present

87. The bigger a clot gets, the slower the blood is lost from a cut vessel, and the slower that blood is lost, the bigger the clot gets:
    (a) Law of Complementarity
    (b) "Fight-or-flight" response
    (c) Negative feedback process
    (d) Trophic hormone influence
    (e) Positive feedback cycle

88. Bronchodilation involves the following general sequence of events:
    (a) Bronchiole narrows, increased resistance, decreased air flow
    (b) Blood vessel narrows, decreased resistance, increased blood flow
    (c) Trachea enlarges, decreased resistance, increased air flow
    (d) Primary bronchus narrows, increased resistance, decreased air flow
    (e) Bronchiole widens, decreased resistance, increased air flow

89. Negative-pressure breathing:
    (a) Artificial respiration
    (b) Shoving a bicycle pump hose between somebody's lips, then pumping like mad!
    (c) Normal, resting expiration
    (d) Using oxygen tanks to breathe while climbing Mount Everest
    (e) Normal, unassisted inspiration

90. The total amount of air a person can inhale and exhale from normal, uncollapsed lungs:
    (a) Total lung capacity
    (b) Residual volume

(c) Tidal volume

(d) Expiratory reserve volume

(e) Vital capacity

91. The enzyme within red blood cells that allows H_2O and CO_2 to combine rapidly:
    (a) Myosin ATPase
    (b) Lactic dehydrogenase
    (c) Carbonic anhydrase
    (d) CO
    (e) Acetyl CoA

92. The major stimulus for inspiration:
    (a) A slight drop in blood $[O_2]$
    (b) A huge increase in blood pH
    (c) Accumulation of carbon monoxide within the alveoli
    (d) A slight increase in blood CO_2 and H^+ concentrations within the blood
    (e) Reduction in alkali level due to vomiting

93. Defecation is the same thing as:
    (a) Ingestion
    (b) Egestion
    (c) Secretion
    (d) Digestion
    (e) Absorption

94. Dietary fiber eaten as whole wheat cereal is:
    (a) Absorbed, but not secreted
    (b) Ingested, but not absorbed
    (c) Digested, but not defecated
    (d) Egested, then ingested
    (e) Digested, then has gestation!

95. Bile is continuously produced and secreted by the:
    (a) Cholecyst
    (b) Pancreatic acinar cells
    (c) Hepatocytes
    (d) Chondrocytes
    (e) Zygotes

96. The major stimulus for gall bladder contraction:
    (a) Increased secretin levels in the blood
    (b) Lower bicarbonate level within the pancreatic juice

(c) Elevated bilirubin concentration inside the bloodstream
(d) A blockage of the common bile duct
(e) Raised blood CCK levels

97. The main poisonous, nitrogen-containing solute excreted in the urine:
    (a) Ammonia
    (b) Sodium chloride
    (c) Penicillin
    (d) Glucose
    (e) Urea

98. About ___% of the ejaculate actually consists of sperm cells:
    (a) 5
    (b) 10
    (c) 25
    (d) 50
    (e) 60

99. The female reproductive pathway begins with the zygote and ends with the:
    (a) Placenta
    (b) Neonate
    (c) Embryo
    (d) Fetus
    (e) Amnion

100. The two primary hormones that act synergistically together to powerfully start labor:
    (a) CCK and GH
    (b) Insulin and glucagon
    (c) Protein and vitamin C
    (d) Oxytocin and prostaglandins
    (e) Oxytocin and ADH

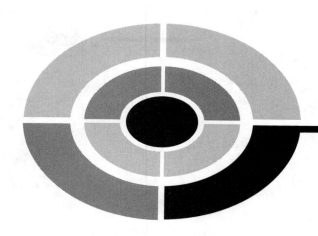

Answers to Quiz, Test, and Exam Questions

Chapter 1

1. B	2. C	3. A	4. D
5. D	6. C	7. D	8. A
9. B	10. A		

Chapter 2

1. C	2. A	3. B	4. D
5. C	6. B	7. A	8. B
9. A	10. B		

Answers

Test: Part 1

1. B	2. D	3. E	4. A
5. B	6. C	7. B	8. A
9. E	10. E	11. C	12. A
13. D	14. A	15. A	16. D
17. B	18. B	19. C	20. E
21. D	22. A	23. C	24. B
25. D			

Chapter 3

1. D	2. B	3. D	4. C
5. B	6. D	7. D	8. B
9. C	10. D		

Chapter 4

1. B	2. C	3. A	4. A
5. D	6. D	7. D	8. A
9. C	10. C		

Test: Part 2

1. C	2. A	3. C	4. C
5. E	6. B	7. E	8. C
9. C	10. A	11. C	12. D
13. C	14. E	15. A	16. C
17. E	18. B	19. A	20. C
21. D	22. C	23. D	24. E
25. B			

Chapter 5

1. A	2. C	3. A	4. D
5. B	6. B	7. A	8. D
9. A	10. D		

Chapter 6

1. B	2. D	3. C	4. D
5. A	6. C	7. C	8. A
9. D	10. C		

Test: Part 3

1. D	2. A	3. A	4. D
5. A	6. C	7. C	8. A
9. E	10. B	11. D	12. A
13. B	14. C	15. A	16. D
17. B	18. C	19. E	20. A
21. B	22. D	23. C	24. B
25. D			

Chapter 7

1. B	2. D	3. A	4. D
5. D	6. A	7. C	8. B
9. C	10. B		

Chapter 8

1. C	2. A	3. D	4. C
5. B	6. C	7. D	8. A
9. B	10. A		

Test: Part 4

1. A	2. C	3. A	4. C
5. A	6. D	7. C	8. D
9. C	10. A	11. B	12. E
13. A	14. C	15. B	16. A
17. C	18. E	19. E	20. E
21. E	22. B	23. C	24. A
25. D			

Chapter 9

1. D	2. A	3. D	4. C
5. C	6. B	7. C	8. D
9. C	10. D		

Chapter 10

1. C	2. A	3. B	4. A
5. C	6. B	7. D	8. B
9. A	10. C		

Chapter 11

1. D	2. D	3. A	4. B
5. B	6. A	7. D	8. C
9. B	10. D		

Test: Part 5

1. C	2. D	3. A	4. A
5. E	6. E	7. B	8. A
9. B	10. A	11. B	12. C
13. C	14. B	15. A	16. B
17. C	18. A	19. E	20. A
21. B	22. D	23. B	24. D
25. E			

Chapter 12

1. B	2. A	3. A	4. C
5. B	6. D	7. D	8. C
9. A	10. C		

Chapter 13

1. C	2. A	3. D	4. A
5. C	6. D	7. B	8. D
9. A	10. D		

Test: Part 6

1. E	2. B	3. A	4. A
5. D	6. C	7. A	8. B
9. A	10. D	11. C	12. A
13. E	14. A	15. A	16. A
17. B	18. B	19. D	20. A
21. D	22. E	23. D	24. E
25. B			

Final Exam

1. C	2. B	3. C	4. A
5. E	6. C	7. D	8. C
9. B	10. C	11. C	12. C
13. E	14. E	15. A	16. B
17. C	18. D	19. B	20. D
21. A	22. C	23. E	24. D
25. A	26. E	27. C	28. B
29. D	30. D	31. A	32. C
33. E	34. D	35. E	36. B
37. B	38. D	39. B	40. E
41. A	42. C	43. A	44. B
45. A	46. C	47. D	48. B
49. D	50. B	51. A	52. B
53. B	54. D	55. A	56. B
57. C	58. E	59. E	60. B
61. B	62. A	63. C	64. E
65. B	66. D	67. B	68. C
69. E	70. A	71. E	72. C
73. B	74. C	75. E	76. E
77. E	78. C	79. B	80. C
81. D	82. D	83. E	84. A
85. C	86. D	87. E	88. E

Answers

89. E	90. E	91. C	92. D
93. B	94. B	95. C	96. E
97. E	98. B	99. B	100. D

INDEX

ABOUT THE AUTHOR

Dale Layman, Ph.D., Grand Ph.D. in Medicine, is a popular professor of biology and human anatomy and physiology at Joliet Junior College in Illinois. He is the author of many articles, as well as *ANATOMY DEMYSTIFIED* and *BIOLOGY DEMYSTIFIED*, two closely related books in this series. He is the first Grand Doctor of Philosophy in Medicine for the United States, and has received many other international honors and awards for his writing, teaching, and highly creative thinking. Dr. Layman has more than 28 years of experience in the biological sciences.